Polyvinyl Alcohol-Based Biocomposites and Bionanocomposites

Scrivener Publishing
100 Cummings Center, Suite 541J
Beverly, MA 01915-6106

Publishers at Scrivener
Martin Scrivener (martin@scrivenerpublishing.com)
Phillip Carmical (pcarmical@scrivenerpublishing.com)

Polyvinyl Alcohol-Based Biocomposites and Bionanocomposites

Edited by

Visakh P. M.

Department of Physical Electronics, TUSUR University, Tomsk, Russia

and

Olga B. Nazarenko

School of Non-Destructive Testing, Tomsk Polytechnic University, Tomsk, Russia

Scrivener Publishing

WILEY

This edition first published 2023 by John Wiley & Sons, Inc., 111 River Street, Hoboken, NJ 07030, USA and Scrivener Publishing LLC, 100 Cummings Center, Suite 541J, Beverly, MA 01915, USA

For more information about Scrivener publications please visit www.scrivenerpublishing.com.

Wiley Global Headquarters
111 River Street, Hoboken, NJ 07030, USA

For details of our global editorial offices, customer services, and more information about Wiley products visit us at www.wiley.com.

Library of Congress Cataloging-in-Publication Data

ISBN 978-1-119-59209-9

Cover image: Pixabay.Com
Cover design by Russell Richardson

Set in size of 11pt and Minion Pro by Manila Typesetting Company, Makati, Philippines

Printed in the USA

10 9 8 7 6 5 4 3 2 1

Contents

Preface

Many of the recent research accomplishments in the area of polyvinyl alcohol (PVA)-based biocomposites and bionanocomposites are summarized in this book. In it, we have tried to discuss as many topics as possible on the most recent state-of-the-art developments regarding these biocomposites and bionanocomposites, the challenges faced when using them, and their future prospects. In addition to providing a biodegradation study of them, their significance and applications are also discussed, along with practical steps towards their commercialization. Moreover, PVA/cellulose-based and PVA/starch-based biocomposites and bionanocomposites are discussed, along with the biomedical applications of PVA-based composites and nanocomposites, and PVA-based hybrid interpolymeric complexes and their applications. As can be seen from the range of topics mentioned above, this book will be a very valuable reference source for university/college faculties, professionals, post-doctoral research fellows, senior graduate students, and researchers in R&D laboratories working in the area of PVA. Since the various chapters are contributed by prominent researchers from industry, academia and government/private research laboratories across the globe, the book can be used as an up-to-date resource on the major findings and observations in the field.

In Chapter 1, an overview of PVA-based biocomposites and bionanocomposites is presented that includes their scope of application, state-of-the-art preparation methods, new challenges and opportunities. Chapter 2 presents a biodegradation study of PVA-based biocomposites and bionanocomposites. In addition to biodegradable PVA biocomposites and bionanocomposites, the authors also discuss many other topics, including biocomposites and bionanocomposites based on PVA/starch, PVA/hemicellulose, PVA/polylactic acid and PVA/polyhydroxyalkanoates. The significance of PVA-based biocomposites and bionanocomposites and their applications are discussed in Chapter 3, along with practical steps to take towards their commercialization. Next, different parts of the chapter discuss the properties of PVA composites and nanocomposites, their categorization and advantages, and other issues associated with them, along with their future prospects.

In the first part of Chapter 4, the authors focus on the preparation of PVA/cellulose-based biocomposites and bionanocomposites. The various topics discussed include PVA/cellulose fibers, PVA/cellulose acetate, PVA/bacterial cellulose, PVA/regenerated cellulose, PVA/cellulose aerogel or hydrogel, PVA/cellulose nanocrystals and PVA/cellulose nanofiber. The second part of the chapter discusses the methods used to characterize them, such as tensile and thermal characterizations, X-ray diffraction, and morphological, rheological and viscoelastic characterizations. In the third part of the chapter, their potential applications are discussed. Next, Chapter 5 provides a good framework for the study of PVA/starch-based biocomposites and bionanocomposites. After a detailed introduction, their preparation, characterization and applications are discussed.

In Chapter 6, the authors discuss PLA/polylactic acid-based composites and bionanocomposites. Included in the discussion is the role of plasticizers and fillers in composite development, the methods employed in the development of structured polymers, the techniques used to analyze them, and their applications. Next, Chapter 7 discusses the biomedical applications of PVA-based bionanocomposites. The authors focus on their application in drug delivery systems, wound healing, tissue engineering and regenerative medicine, and also discuss their future perspectives. The book concludes with Chapter 8, which is a detailed introduction to hybrid interpolymeric complexes, in which their production and possible applications are also discussed.

Finally, we would like to express our sincere gratitude to all the contributors to this book, whose excellent support and enthusiasm has led to the successful completion of this venture. We are grateful to them for the commitment and sincerity they showed towards their contributions. We would also like to thank all the reviewers for using their valuable time to make critical comments on each chapter. We also thank Scrivener Publishing for recognizing the demand for a book on the increasingly important area of PVA-based biocomposites and bionanocomposites, and for their interest in publishing a book on subjects which have yet to be addressed by many other publishers.

Dr. Visakh P. M.
Dr. Olga Nazarenko
February 2023

1

Polyvinyl Alcohol-Based Biocomposites and Bionanocomposites: State-of-the-Art, New Challenges and Opportunities

Visakh P. M.

Department of Physical Electronics, TUSUR University, Tomsk, Russia

Abstract

This chapter presents the recent advances in the field of polyvinyl alcohol-based biocomposites and bionanocomposites and their new challenges and opportunities. In this chapter, we will be discussing mainly short abstract for all chapters in this book, with different topics, such as biodegradation study of polyvinyl alcohol-based biocomposites and bionanocomposites, polyvinyl alcohol-based biocomposites and bionanocomposites: significance and applications, practical step toward commercialization, polyvinyl alcohol/cellulose-based biocomposites and bionanocomposites, polyvinyl alcohol/starch-based biocomposites and bionanocomposites, polyvinyl alcohol/polylactic acid–based biocomposites and bionanocomposites, biomedical applications of polyvinyl alcohol-based bionanocomposites and hybrid interpolymeric complexes.

Keywords: Polyvinyl alcohol, biocomposite, bionanocomposites, biodegradation, nanocomposites, hybrid interpolymeric complexes, biomaterials

1.1 Biodegradation Study of Polyvinyl Alcohol-Based Biocomposites and Bionanocomposites

PVA applications cover the research areas of formulation films, synthesis of coatings, adhesives products, and emulsion polymerization. Globally, PVA production and consumption was assessed nearly 1.124 million tons

Email: visagam143@gmail.com

Visakh P. M. and Olga B. Nazarenko (eds.) Polyvinyl Alcohol-Based Biocomposites and Bionanocomposites, (1–30) © 2023 Scrivener Publishing LLC

in 2016. Polyvinyl alcohol (PVA) exhibits the properties, such as thermostability, water solubility, film forming, high viscosity, emulsifying, tensile strength, and flexibility [1]. In biodeterioration, the microbial biofilm populates the surface of substrate on which they initiate apparent biodegradation. This changes the morphology of the surface into more rough and deformed. Later, deplolymerization involved extracellular enzymes, which are secreted by microbial cells. Enzymes catalyze the breakdown of bonds in polymers and produce low molecular weight products like oligomers, dimers, or monomers. Meanwhile, physical and chemical degradation has several disadvantages, such as incomplete decay efficiency, higher cost and by-product pollution [2]. In comparison, microbial and enzymatic degradation is drawing increased attentiveness because of high efficiency, low cost, and more economic and environmental protection compared with physical-chemical degradation [3]. Various microorganisms have been found useful for biodegradation of PVA. Diversity of PVA biodegraders has been cited in literature, which spans from natural source, like activated sludge, soil, and biodegradable support, like polymeric sheet. These strains of microorganisms have shown their ability to efficiently assimilate PVA as carbon source in growth medium. These microorganisms are studied either as a pure culture or mixed culture to demonstrate better activity on PVA [4].

A few strains of bacterial and fungus that have been utilized for PVA biodegradation reported are *Pseudomonas, Alcaligenes,* and *Bacillus. Penicillium* WSH02-21, *Actinomycete, Streptomyces venezuelae* GY1 strain. *Aspergillus foetidus.* The expression of the PVA enzymes can be inducible under appropriate conditions. Bacterial species that utilize PVA have been found from sludge samples by providing PVA as a selective source of carbon. PVAase, an enzyme that degrades PVA, secreted from *Bacillus niacin* immobilized by cross-linking as enzyme aggregates has shown improved enzyme activity approximately 90% compared with free PVAase. Non-purified PVAase can increase the usability for its large-scale application in the industry [5]. Recyclable biocomposites derived through naturally degradable polymers are prepared to make composition attractive. This strategy enhances biocomposites properties after blending with another nanosize biodegradable filler material for better processability, usability and extending life of end-use application [6]. The orange peel powder-improved properties of PVA films and make the PVA suitable for packaging application [7]. The biodegradability feature of PVA nanocomposite films lessened in a certain degree by modification with expensive inorganic nanoparticles of graphene oxide nanosheet [8], calcium carbonate nanoparticles [9], ZnO and nano-SiO_2 [10]. These nanocomposite films have showed significant improvements in the barrier performance due to presence of nanofillers.

Biodegradability of PVA/cellulose composites functioned at 22°C to 27°C and relative humidity ranges 70% to 80%. Samples had displayed a fast weight loss in 16 days in a soil burial test. Weight loss has decelerated in the succeeding soil burial period after 16 days. Their work concludes that the cellulose biodegradability rates are higher than PVA in ecocomposites which resulted in higher weight loss with better biodegradability than that of neat PVA. In biodegraded biocomposites, recovery and analysis of constituent make the composites from complex matrix, like soil compromised the study outcome. For example, cellulose fiber collection from soil after the first 16 days of biodegradation of PVA/cellulose in a buried soil severely hinders degree of biodegradation under natural decomposing conditions. PVA mixed with cellulose prepared through 40 cycles of pan milling was more possible to biodegrade than cellulose obtained through single cycle of pan milling. Higher number of cycles of pan milling reduces the size of cellulose fibers <20 μm and less crystalline, which promote the substrate utilization activity for microorganisms through enhanced substrate-microorganism interaction in the soil [11].

The composition of PVA/nanowhiskers biocomposites discloses its biodegradable and environmental friendly biomaterials with utilization as a food source after disposal in a manner that has a positive effect during natural degradability. Strengthening of poly(vinyl alcohol) nanocomposites after the addition of alpha-chitin nanowhiskers has shown improved mechanical properties [12], which possibly increased the resistance to biodegradation. Machine-driven reinforcement of PVA tricomponent nanocomposites with chitin nanofibers and cellulose nanocrystals known for better thermal properties [13], which reduces the biodegradability under ambient temperate conditions. In fact, tricomponent nanocomposites change their mechanical and thermal characteristics that turn them into high-performance biomaterials with possible low environmental impact. The PVA/CS film amalgamated with various concentrations of CNC, which were 1, 3, and 5 wt%. The PVA/CS/CNC bionanocomposite film demonstrated excellent response as antifungal and antibacterial activity, a property which is mandatory and associated with potential films for food processing industry [14]. PVA has special interest in biocomposites as it has the ability to reduce antioxidant and antibacterial activities against gram-positive and gram-negative bacteria tested [15]. The mixing of PVA into chitosan has remarkably accepted as a new way to obtain films with promising biodegradability while retaining reasonable antioxidant and antibacterial properties. It provides balanced biocomposites with good strength, biodegradability, and application in extending the shelf life of packed food.

1.2 Polyvinyl Alcohol-Based Biocomposites and Bionanocomposites: Significance and Applications, Practical Step Toward Commercialization

Key raw material to prepare PVA is the vinyl acetate monomer. The monomer is manufactured through the polymerization of vinyl acetate. Instead, it goes through partial hydrolysis, which consists of partial substitution of the ester group with the hydroxyl group in vinyl acetate, completed in the presence of aqueous sodium hydroxide. The PVA is precipitated, washed, and dried after gradual application of the aqueous saponification agent. When making the polyvinyl alcohol solution, it is recommended to use tap water, as bacteria grow faster in PVA containing distilled water. This allowed macromolecules to form crystallites, stabilizing the films and inducing a chemically cross-linked behavior. It has outstanding optical properties, great dielectric power, and excellent capacity for storing charges [16]. Doping with nanofillers can readily customize its mechanical, optical, and electrical attributes. Hermann and Haehnel first synthesized it in 1924 by saponifying the poly(vinyl ester) with a solution of sodium hydroxide resulting in a PVA solution [17]. PVA's physicochemical and mechanical properties are governed by the number of hydroxyl groups contained in the polymer PVA [6]. Different grades of PVA are available on the market based on hydrolysis (percent) and molecular mass and have different characteristics, including melting point, viscosity, pH, refractive index, and band gap [18]. The consequence of variation of the length and the degree of hydrolysis of vinyl acetate under acidic or alkaline conditions results in different PVAs having different durability, tensile strength, density, emulsification extent, dispersing capacity, etc.

PVA has a high melting point, due to hydrogen bonding in the matrix. Interfaces play a crucial role in understanding material behavior, such as PVA. One downside to dealing with bulk materials is the presence of a small fraction of atoms at the interfacial surface. One major problem in the manufacture of polymeric nanocomposites is the uniform dispersion in polymer matrix of nanofillers. Uniform dispersion plays a crucial role in producing multifunctional composites, and compounding techniques may be used to accomplish this. Compounding involves combining different materials, and such materials may either be a combination of polymers or polymer polymer additives. A polymer composite consists of filler reinforcement and matrix of polymers. The fillers measurements can be in micrometers or nanometers. Because of the hydrophilic nature of PVA, crosslinkers are used to synthesize hydrogels for a number of applications. PVA is extensively used in biomedical applications for its compatibility, PVA composites, such as PVA gels, are used in diverse biomedical fields, such as in the engineering of contact lenses [19], artificial heart surgery [20], drug delivery systems [21], and wound dressings [22].

PVA may be used in a wide range of uses, such as molecular sensing in biological and biomedical fields, membranes of fuel cells, chemo sensors, absorption of toxic metals, and optoelectronic devices. Nanofillers expose their larger surface area for interaction with polymer that is the important idea in the development of useful properties of polymeric nanomaterials. In medical devices, PVA is used as a biomaterial because of its extremely promising properties, for example, biocompatibility, nontoxicity, non-carcinogenic, swelling properties, and bioadhesive features. This material is very valuable and desirable for biomedical application and uses. Excellent mechanical properties of the PVA nanocomposites are intended to be composted at the end of their life rather than end up in landfills like most traditional petroleum-based non-biodegradable plastics. Fabrication and characterization of PVA/TiO_2 nanocomposite films [23] and PVA/amino acid composite [24] for orthopedic applications are demonstrated.

Also, several investigators have documented PVA's wear features. The physical mixture of a polymer with the layered silicate may not form a nanocomposite. This situation is similar to polymer blends, and in most cases, separation into discrete phases may take place. In immiscible systems, which typically correspond to the more conventionally filled polymers, the poor physical interaction between the organic and inorganic components leads to poor mechanical and thermal properties. PVA-based composites/biocomposites are well explored for various applications for energy generating devices and energy storing devices as well. They are capable of improving processability and/or flexibility of the materials and also generate fascinating interfacial properties with innovative functions. Up to the present time, various PVA-based composites/biocomposites with insulating or conducting polymers/carbon nanostructures have been prepared through noncovalent or covalent modifications. Due to their broad applications in high-strength and conductive materials, catalysts, and energy-related systems, graphene/polymer composites have attracted considerable interest, especially flexible energy conversion and storage devices. Recently, a simplistic and scalable methodology to fabricate novel ABC-type terpolymer-based proton-conducting membranes are prepared and elucidated the benefits of utilizing terpolymer composite membrane as an electrolyte for direct methanol fuel cells [25].

These PVA-TAF terpolymer composite membranes showed good thermal stability, flexibility, water uptake, and retention capacity. The prepared terpolymer composite membranes were thermally stable in a dry nitrogen atmosphere. Novel PVA-alginate–based mixed-matrix membranes with heteropolyacids such as phosphomolybdic acid phosphotungstic acid and silicotungstic acid are reported for use as electrolytes in direct methanol fuel cells [26]. Hybrid nanocomposite PVA membranes prepared by integrating amino acid functionalized titanium dioxide biohybrid nanoparticles into a PVA matrix is also studied [27]. PVA/amino acid biocomposite membranes are explored as a new class of biocomposite membrane electrolytes for direct methanol fuel cells by Suganthi et al [28]. Followed by this, Pectin blended with PVA to fabricate new class of hybrid nanocomposite followed by addition of sulfonated titanium dioxide (s-TiO_2) nanoparticles as inorganic proton conducting material [29]. PVA-based electrolyte acted as a prototype of biofunctionalized membranes, which rely on explanatory association between well-organized proton-based biological energy transformation process and proton conduction process in PEFC electrolyte under low humid operating [30]. For environmentally sustainable processing, the combination of synthetic and natural polymers is favored. Synthetic polymers usually have a stable relative to natural polymers.

1.3 Polyvinyl Alcohol/Cellulose-Based Biocomposites and Bionanocomposites

The types of the cellulose can be differentiating based on the configuration, chemical composition and the properties of the cellulose. The classes of the cellulose including cellulose fiber, cellulose nanofiber, cellulose acetate, bacterial cellulose, regenerated cellulose, cellulose nanocrystalline, and cellulose hydrogel. The natural fibers that have higher percentage of cellulose poses excellent mechanical strength but easy to be more flammable. This involvedness occur due to the nature of the cellulose which having strong intermolecular hydrogen bonding, poses a high degree of polymerization and also high crystallinity degree [31]. Conjointly with the environmentally safe guideline, the biodegradable polymer, which is polyvinyl alcohol (PVA), preferably reinforced to cellulose-based materials to create eco-sustainable composites [32]. The fabrication of PVA/cellulose acetate membrane can be completed via common phase inversion method [33, 34] and casting method [35]. Being different, the preparation of CA membrane coated with PVA and Fe_3O_4 nanocomposites done via *in situ* formation of Fe_3O_4 in the polymeric solution containing PVA through facile chemical procedure [36]. The produced nanocomposites PVA/Fe_3O_4 prepared by dissolving PVA in water and *in situ* formation of magnetite nanoparticles with a certain addition amount of Fe_3O_4. Therefore, the prepared CA membrane prepared via phase inversion method was immersed in PVA/Fe_3O_4 before being dried for 2 hours at 60°C. Bacterial cellulose labeled as BC has been considered as a biomaterial with a dazzling future in many industrial fields especially in medical due to its biocompatibility, biodegradability, high clarity, having fine fiber network structure, high surface area, high crystallinity with high holding water capacity and possess excellent strength-to-weight ratio [37–43].

The development of the BC is swiftly observed to widen the application of BC in industrial's demand. As it progresses, high interest was also paid to lessen the BC production costs by substituting with cheaper materials, industrial waste, or by-products. The production of cellulose by aerobic bacteria as an extracellular polysaccharide membrane is a bottom-up approach that finally produced highly pure cellulose. In contrast, the top-down process in the production of cellulose involving the extraction of cellulose sourced from plant and woods. The effect of sonication steps in the fabrication of PVA/starch blended with BC was reported by Abral and groups [44]. Ultrasonication method able to minimize the viscosity of biopolymer and improved the filler dispersion in the matrix [45–47]. Compared to unsonicated PVA/BC, the sonicated PVA/starch blended with BC produced less viscous with the starch gel became shorter and higher in mobility resulting more homogeneous structure of the blend. In another approach, the *in-situ* growth process was done by the direct addition of PVA into *Acetobacter xylinum* inoculated medium and compared with the composites done by impregnation of BC gels with PVA [48].

The formation of film is casted on the glass plate before being coagulated in the distilled water. The transparency of the fabricated PVA/RC compared with RC is demonstrated. As observed, the wrinkle of RC film is obviously compared with PVA/RC film, indicating that the shrinkage of RC greatly improved compared with the RC incorporated with PVA, which is consistent in the mechanical properties. Similar work done by Lu and co-workers in which the dissolution and the blending of PVA/RC are derived from microcrystalline cellulose (MCC) prepared in AMIMCI [49, 50]. Interestingly, the transparent film made up from RC derived from Kenaf core powder that is dissolved in LiOH/urea solvent reinforced with PVA successfully fabricated and can be reprinted repeatedly several times [51]. Briefly, regenerated CNC extracted from pineapple peel is first dissolved in 1-butyl-3-methylimidazolium chloride (BmimCl) ionic liquid solution to RC suspension. The suspension mixed with PVA in cylindrical mold, followed by five freeze-thaw cycle done at freezing condition (–20°C for 8 hours) and thawing for 4 hours at room temperature to form PVA/CNC hydrogel. Additionally, fabricated PVA/CNC hydrogel is magnetized with the Fe_3O_4. The preparation process of the PVA/CNC hydrogel with the addition of magnetic nanoparticles. Recently, by using the same freezing/thawing method, tricaroxy cellulose (OxC) incorporated with PVA [52].

OxC mixture composites and pure PVA aqueous solution are subjected to three freezing thawing cycles. After each of the freezing cycle, the OxC/PVA kept at 4°C to ensure the formation of porous structure. Next, the preparation of PVA/CNC/poly(2-Hyroxyethyl methacrylate) (PVA/CNC/polyHEMA) and PVA/CNC/poly(N-methylenebisacrylamide) (PVA/CNC/polyMBA) hydrogels prepared by photo-crosslinking followed by freezing/thawing cycle [53]. As proposed by Tien Lam and groups, the compatibility of the produced PVA/CNC with human fibroblast skin line demonstrated a good spreading and adhesion of cells on the materials surfaces [54]. The PVA/CNC shows the uniform, porous morphology without any shrinkage after the cell growth.

1.4 Polyvinyl Alcohol/Starch-Based Biocomposites and Bionanocomposites

These days, polyvinyl alcohol (PVA) is the most significant plastic materials used in the industrial among biodegradable polymers [55]. It is an ecologically biodegradable polymer with excellent mechanical properties [56]. Several studies have been reported on water-soluble PVA, where it has been found to be mineralized and biodegraded in numerous environments condition [57]. The application of PVA is mostly implemented in agricultural mulch films, paper coatings, and packaging application. Besides, PVA is also used as lubricant to make biodegradable carriers or water soluble in eye drops and hard contact lens solution [58]. The biodegradability of PVA in various microbial environments has been widely reported [59–64]. Starch is one of the best option to be blended with PVA [65, 66]. Starch is made up of two types of materials, amylose, a linear chain molecule of α-1,4-linked D-glucose and amylopectin, a branched polymer of α-1,4-linked D-glucose with a 1,6-linked D-glucose unit. It is reflected as a condensation polymer of glucose. In fact, it is capable to be biodegraded by a number of enzyme through enzyme-catalyzed hydrolysis [67]. The PVA/starch blends have shown an excellent compatibility [68, 69]. In fact, PVA and starch exhibited highly desirable physical blend in various applications. Blend of these PVA-based materials with starches permits the manufacturing of water-soluble blends at a lower cost while maintaining the high properties and characteristics of PVA. Several studies have reported their patented formulation of PVA/starch blend showed an enhancement in strength, modulus, as well as compatibility of the blend. However, their tensile strength, elongation at break, as well as their water resistance properties, need to further increase [70, 71].

The compatibility between starch and synthetic polymers can be improved by the addition of compatibilizer or chemical modification [72–75]. Apart from that, cross-linking, etherification, esterification, and oxidation also can be applied for starch modification. Above all, methylation was usually chosen as one of the esterification method in interpreting the substitution pattern in polymer chains and the structure of polysaccharides. They have been verified as an effective technique to develop the blend's properties [76]. In another way, crosslinking agent also can be added to improve the physical and mechanical properties of the starch/PVA blends [77]. PVA/starch blend was also prepared by Susmita Dey Sadhu *et al.* [78], where the PVA/starch mixture was first heated at 70°C until uniformity appears. Then, the solutions were cast on a casting mold and dried in an oven at 75°C. Another blend preparation was reported by Zhijun Wu *et al.* [79] on PVA/starch blend.

First, the PVA, starch and glycerol were dissolved in distilled water and the blend was stirred using an electric stirrer for 45 min at 95°C. Next, the solutions were poured onto the glass plates and dried for 24 hours at room temperature. The dried films were removed from the glass plates and kept for further testing. Mechanical properties define a material's behavior once subjected to mechanical stresses. The properties include tensile strength, elongation at break, and tensile modulus measurements. As reported by Zaaba *et al.* [80], the tensile strength (TS), elongation at break (Eb) and tensile modulus of PVA/tapioca starch biodegradable films were evaluated according to ASTM D882 using the Instron 3366 testing machine. Each film was cut with about 0.09-mm average thickness and 6.4-mm width. In another study, Shangwen Thang *et al.* [81] reported that the tensile strength and elongation at break of the PVA/starch/nano-silicon dioxide (nano-SiO_2) biodegradable blend films were measured according to Chinese standard method GB/T4456-96 (Polyethylene Blown film for packaging, 1996) using the electron tensile tester CMT-6104.

Shangwen Thang *et al.* [82] reported the tensile strength and elongation at break of PVA/starch/nano-SiO_2 biodegradable blend films at different nano-SiO_2 content. The tensile strength of PVA/starch without nano-SiO_2 was 9.03 MPa. As the content of nano-SiO_2 increase, the tensile strength of the blend also increased (15.0 MPa), up to touched at maximum point which was about 2.5 wt.% of the nano-SiO_2 content. After that point, the tensile strength of the blend film started to decrease along with the increment of nano-SiO_2 content. Zhijun Wu *et al.* [53] stated in their findings on the TGA of starch/polyvinyl alcohol/citric acid ternary (S/P/C) blend. As can be seen from the thermogravimetric curves, the weight decreased along with increasing temperature while the DTGA curves displayed the maximum decomposition temperature (T max) of thermal decomposition [83].

1.5 Polyvinyl Alcohol/Polylactic Acid-Based Biocomposites and Bionanocomposites

Polymers find a variety of applications in medicine ranging from the development of heart valves, blood vessels, urinary catheters, artificial skin and kidney to hemodialysis membranes. They have found applications in the nanobased drug delivery systems, where, the biodegradable polymers are exploited. Recently, both the treatment and imaging techniques make use of PLGA (poly lactic-co-glycolic acid), PGA (poly glycolic acid) and PLA (poly lactic acid) which has been approved by the FDA for the medicinal usage. The polymers, such as poly(ethylene terephthalate), polyurethanes, microporous silicon rubber, find its application in the vascular prosthesis, whereas polypropylene is used in cardiopulmonary bypass surgery. The semipermeable membranes of cellulose have been exploited in the hemodialysis. The mucoadhesive polymers are used in the ophthalmic and the buccal delivery of the drugs. Polyacrylic acid (PAAc) is a common bioadhesive drug delivery system. PVA can be in either of the two forms, partially hydrolyzed or fully hydrolyzed. The variations in the solubility, flexibility, molecular weight adhesiveness, and tensile strength of the PVA can be attributed to the variations in the length of both vinyl acetate polymer and the degree of hydrolysis [84]. PVA found its first application in the rayon textiles industry as a wrap sizing material along with its applications in the paper coating [85]. It is easily biodegradable and maintains its crystalline structure in water. It has found its applications in almost all the sectors ranging from industrial, medicinal to commercial. Other common applications of PVA include the synthesis of the biodegradable protective apparel, paper products, tube winding, carton sealing, thickening agent for latex paint, common household glue, and in other adhesive mixtures.

Poly lactic acid is a thermoplastic aliphatic polyester, which can be derived from renewable sources, such as corn starch, sugarcane, and tapioca roots. PLA remains one of the most promising polymers considering its properties such as biocompatibility, biodegradability, ready availability and low cost, in addition to its outstanding processing performance. However, brittleness and poor thermal stability remain as the limiting factors [86]. The copolymerization of the PLA or the introduction of plasticizer has thus become a suitable approach to tune its characteristics. The modification of the PLA can be bought about through the employment of various techniques, such as copolymerization, crosslinking, and blend formation, which in turn can lead to the development of tailor-made material. However, it has been proposed that better toughening results could be obtained through the incorporation of nanoparticles as fillers. According to Hoidy *et al.* [87], the mechanical properties of the PLA could be improved by the incorporation of MMT clay.

PLA films have found its application in both medicine and packaging and are being considered as an alternative to the synthetic packaging materials. Large and sharper peaks can be obtained with an increment in the PVA concentration. However, this endothermic peak was softened with the addition of the plasticizer, glycerol. This could be because of the interaction of the PVA with glycerol, resulting in corresponding film to be thermally stable. The thermal degradation of the cellulose occurs between 225 and 325°C [88]. The intelligent pH sensing wraps developed using PVA/starch complex having incorporated zinc oxide nanoparticles and phytochemical constituents is a novel type of active packaging. Among the various packaging materials available the antimicrobial packaging is a common application. Here, the matrix of the material is incorporated with antimicrobial components, which can either be a nanomaterial or micromaterial or an herbal composition. The *in vitro* susceptibility of the bacteria to the active components present in the packaging material determines the antimicrobial potential of the film.

The polymers find a wide range of applications in medicine. They are being exploited as the drug carriers for specific targeted delivery, as well as the sustainable release of the drug. The ophthalmic composition to the surgical interventions includes a wide range of polymers. Hydrogels are defined as gels containing water but not soluble. PVA hydrogels have found its application as an artificial vascular material replacement considering its low toxicity, excellent mechanical properties comparable to the vascular system and biocompatibility. The freeze thaw method employed in the PVA hydrogel preparation can overcome the issue of turbidity. The method employs heating a mixture of PVA/water/DMSO (dimethyl sulfoxide), agitating under nitrogen current and then cooling the solution at temperature as low as −10°C for 10 hours.

The biologically decomposable polymers originate from petroleum-based synthetic materials, which decompose naturally under aerobic or anaerobic conditions. The drastic increment in the PVA utilization has led to the raise of substance in the water bodies across the globe after being used in the textile and paper industries. Its foam-forming ability prevents the recovery of oxygen in the so-polluted water bodies. Microorganisms ubiquitous to septic systems, landfills, composts, and soil have been found to be able to degrade the PVA enzymatically. The oxidases and hydrolases when combined can convert the PVA into acetic acid. However, the rate of biodegradation is dependent on the percent hydrolysis and the solubility of the molecule. The blending of the PLA can alter the degradation profile of the composite even though the degradation profile remains identical to that of the parent polymer, the rate and the level depends on the composition and morphology of the individual components [89].

Radiation-induced degradation implies on using the reactive species generated through the radiolysis of water to oxidize PVA [90]. However, coast remains a major concern with the wide application of this method for PVA degradation. Although considering the photocatalytic degradation, the specific conditions required limits its applications [91]. Again, PVA with higher degree of hydrolysis degraded faster than that with lower hydrolysis. This could be due to the higher number of hydroxyl groups which induced higher number of chain scissions, which in turn made the degradation easier by reducing the chain size [92]. The degradation mechanism of the polymer can be varied with the applied temperature where higher degradation can be resulted from a higher temperature.

1.6 Biomedical Applications of Polyvinyl Alcohol-Based Bionanocomposites

Biocompatible nanocomposites can be formed using matrix biocompatible polymers, such as polyvinyl alcohol (PVA) [93]. PVA is a synthetic biodegradable polymer that can mimic natural polymers, due to its characteristics of biocompatibility in human tissues, high hydrophilicity, nontoxicity, corrosion resistance, mechanical flexibility, excellent optical transmission, water solubility, and nonmutagenic behavior [94–98]. Additionally, PVA properties can also be enhanced or changed by the introduction of nanofillers in this polymer. The employed technique is also an important parameter to improve target properties, and some of these techniques include: electrospinning [99, 100], solvent casting [101], metal ion coordination [102], freeze-casting followed by freeze-drying [103], lyophilization [104, 105], thermal cycles [106], ultrasonication process [107], and salt leaching [108]. Nanocomposites using PVA have many applications in the biomedical field, including drug delivery systems [109–111], wound healing, tissue engineering [112], and regenerative medicine. When compared with pure PVA, PVA nanocomposites may improve the protection against bacteria, thermal stability [113, 114], mechanical properties [115], and bioactivity. This chapter will focus on design and challenges of nanocomposites using PVA as a matrix in biomedical applications. Many drugs lose or have their effect diminished due to degradation, and this can occur even before they reach the correct destination/path [116]. Therefore, it is necessary to develop a methodology that delivers these drugs while preserving their pharmacological properties during the journey inside the body, as well as to avoid the release in areas that are not within their target destination. Another important factor is the administration of drugs level within the bloodstream and commonly, as soon as the drug is administered, the initial concentration of the drug in the body is high and this accumulation decreases quickly over time — burst release. Consequently, after a certain period, this drug is no longer effective and is one of the main challenges in this field [117, 118].

Graphene oxide (GO) and its associated compounds exhibit a great potential to drug carrier in cancer treatments [119]. In a PVA nanocomposite filled with GO, a pH sensitivity was detected in drug release process [120]. Superparamagnetic iron oxide nanoparticles (SPIONs) are biocompatible, nontoxic, and biodegradable [121, 122]. The sensitivity of magnetic nanoparticles to the external magnetic field are useful to reach specific biological tissue targets conducting the drugs using magnetic forces. PVA silver nanocomposites can keep the metal release under control. Using silver nanoparticles with carbon nanotubes as PVA nanofillers, the antibacterial behavior can be extended up to 72 hours [123]. The optical properties of PVA/silver nanoparticles can be used to create a transparent wound dressing, which maintain visible the wound to medical diagnose with the injury free from bacteria for longer periods [124]. Alginate is a polymer used in pharmaceutical drugs, wound healing, and when mixed with PVA present nonirritation effect on the eye and skin tissue [125]. Calcium alginate PVA nanocomposites prepared by electrospinning presented in rat wound, healing capacity and showed a direct relationship between the calcium alginate amount and the antibacterial activity against to *Staphylococcus aureus* [126]. A porous nanocomposite made combining Na alginate, bioglass, and PVA prepared by electrospinning technique displayed improved mechanical properties and bioactivity when compared with PVA nanofibers.

Mechanical properties of PVA nanocomposites can be enhanced by filling with nanoclay [127, 128], nanoclay into PVA matrices has demonstrated improvements, such as less flammability [129]. The tensile modulus, hardness, and elongation at break presented enhancements increasing the nanoclay percentage in the PVA matrix hydrogel nanocomposites prepared by freeze-thaw method]. In addition, 3-mm PVA-nanoclay can keep the environment sterile for one week. Comparatively, even more than 60 gauze sheets have this level of protection against bacteria. A natural Na-montmorillonite (hydrophilic nanoclay) mixed with PVA showed an inverse relation between of the nanoclay percentage amount and dehydration rates, and a direct relation with the temperature [130]. In addition, low amounts of Na-rich Montmorillonite in PVA matrix exhibited better tensile modulus and elongation than pure PVA [131]. This nanofiller also enhanced mechanical properties, swelling performance, biocompatibility, antibacterial activity, and did not affect the A-431 cell line (human skin cell) viability blended with PVA/Chitosan nanocomposites [131].

Bacterial nanocellulose can cause enhancements when forming nanocomposites with PVA. These improved characteristics are mechanical properties, water permeability, suggesting good application on blood vessels scaffolds [132] and articular cartilages. An interesting relation was found between filler of cellulose and hydroxyapatite: while cellulose nanofibers decelerate the degradation process, the nanoparticles of hydroxyapatite cause the inverse effect. Fillers of zirconium phosphate (doped with Ca, Mg, and Ti) nanoparticles promoted, also, improvements on mechanical properties, controlled degradation, and bioactivity superior to pure PVA [133]. Titanium dioxide TiO_2 (3%wt.) nanosized, by hydrogen bonds interaction with PVA when mixed with PVA, presented enhancements on mechanical, thermo-stability, and osteoblast cell adhesion. In addition, there are many reports on the use of PVA matrix polymer with several nanocomposites for this specific field, such as nano-chitosan [134], -hydroxyapatite [135] and halloysite nanotubes [136]; suggesting that these PVA-based nanocomposites can further improve the cell regeneration with higher cellular activity for scaffolds with nanofillers. In addition, the different cellular structure and layers, such as cartilage requires materials to have a multiplayer structure to mimic the properties of these various zones. Reports on multilayered hydrogels with nanocomposites for mimicking tissues with various layers, such as cartilage, exhibits an improvement in cell proliferation and a more homogeneous matrix formation [137].

1.7 Hybrid Interpolymeric Complexes

The existence of HICs as a new class of materials was predicted for the first time in 2013 [138]. Earlier, only two representatives of this class were discussed. One of them is a complex of copper hydroxide with polyvinyl alcohol (PVA) (-CH_2-CHOH-)n [139]. Then, well-known complex of starch or PVA with iodine also can be attributed to such materials. It is supposed that in this complex iodine molecules I_2 form a linear chain inside a helix of polysaccharide or PVA. Within iodine chain I_2 single molecules apparently can be still recognized. Different structures of this complex are also discussed in researches [140]. It has been suggested above that HICs can be regarded as a special case of complexes of inorganic molecules with complexing organic polymers. In other words, they are relative materials with some quite similar properties. Other materials relative to HICs can also be pointed. Among them are coordinative polymers (CPs) in which repeated units linked by coordinative bonds can be identified. Abovementioned MOFs are the most known kind of CPs. Hybrid crystalline compounds represent other sort of CPs [141, 142]. Now, MOFs are subjects of numerous researches due to peculiarities of their structure that gives rich possibilities for design of functional materials based on them. The main distinctive feature of MOFs is the presence of pores and channels that makes them suitable for different molecules intercalation. MOFs are also called as "porous coordinative polymers" (PCPs). Other hybrid crystalline compounds have no such pores and channels. At the first stage, soluble zinc salts, like chloride or acetate, are usually used [143–146]. The final aim is obtaining of ZnO or ZnS species prospective for applications. It is known that solvated zinc ions readily bound by complexing polymers like PVA [147].

A type of anion counts for following processing. It is desirable that anions could be easily extracted from the system. Chlorides are accessible and have simple but quite hardly removable anion. A typical second step consists of interaction of zinc salt with ammonium water solution in accordance with expression. Obtained ammonium chloride should be eliminated from the system. It can be achieved by thorough washing. Using of water ammonia solution in step 2 of PVA-$Zn(OH)_2$ HIC production is not a good idea because PVA absorbs water. It leads to diffusion hindrance elimination and formation of composite instead of HIC. As alternative ones can suggest treatment of solid specimen in ammonia vapor or in alcohol alkaline solution. Alcohol is not absorbed well by PVA but use of alkaline needs remove reaction byproduct. The similar problem appears at trying to produce HICs with chalcogenides. For example, if we are going to synthesize PVA-CdS HIC, we can use treatment of specimen in gaseous H_2S or in alcoholic solution of Na_2S. The first method was used in for PVA-CdS nanocomposites production. The second way was used in Prosanov *et al.* [148]. NaCl was produced there as byproduct, which is hardly removable from the system. Rapid complex precipitation can prevent the absorption of boric acid molecules on PVA in sufficient for HIC formation quantities. In our research we used 1% PVA water solution to avoid this problem [149].

It is commonly adopted that separate globules of PVA macromolecules exist in solution with lower than 3% concentrations. Therefore, at lower concentrations interchains crosslinking can be excluded. However, intrachain crosslinking still remains. Such crosslinking complicates modeling sufficiently. In our research, we neglected by this effect for the sake of computation model simplicity. HIC specimens look like brittle opaque films. At overstoichiometric content of boric acid, its explicit precipitation becomes notable. It seems that oversaturated mixture tends to decay on HIC and boric acid instead of composition formation. PVA with copper hydroxide was the first substance, which was suggested as genuine HIC forming compound. PVA-copper hydroxide HIC can be obtained directly from ammonia solution, excluding predecessor conversion step. As it was mentioned in Introduction, PVA-copper oxide HIC formation is expected as a result of PVA-copper hydroxide HIC thermal decomposition. The latter is a subject of great interest due to diverse applications of copper oxide and it was a reason for investigating of these systems in Yu *et al.* [150]. Beneath, we will resume CuO (not Cu_2O) as copper oxide. Complexes of PVA with hydroxides and oxides of Be, Mg, Zn, Cd, Al, Cr(III), and Fe(III) were investigated in. the complex of PVA with boric acid H_3BO_3 (which also can be regarded as hydroxide $B(OH)_3$) was investigated there as well. Producing of this complex excludes preliminary stage. It is the main reason why it was discussed separately. Data concerning complexes of PVA with titanic acid/titanium dioxide and with stannic acid/stannic dioxide were represented in [151] and [152] correspondently. Stannic acid H_4SnO_4

and titanic acid H_4TiO_4 also can be regarded as hydroxides $Sn(OH)_4$ and $Ti(OH)_4$. Actually, there are a number of relative compounds, like H_2SnO_3, $H_{10}Sn_5O_{15}$, H_3TiO_3, and so on. Their complexes have similarities with complexes of other hydroxides, which sometimes can be regarded as hydrated oxides. It is why they are discussed in this section.

References

1. Kawai, F. and Hu, X., Biochemistry of microbial polyvinyl alcohol degradation. *Appl. Microbial. Biotechnol.*, 84, 2, 227, 2009.
2. Bing, X. and Yatong, X., Practice of treating PVA-bearing desizing wastewater. *Environ. Eng.*, 5, 7–9, 2002.
3. Du, G., Liu, L., Song, Z., Hua, Z., Zhu, Y., Chen, J., Production of polyvinyl alcohol-degrading enzyme with Janthinobacterium sp. and its application in cotton fabric desizing. *Biotechnol. J.: Healthcare Nutr. Technol.*, 2, 6, 752–758, 2007.
4. Chiellini, E., Corti, A., D'Antone, S., Solaro, R., Biodegradation of poly(vinyl alcohol) based materials. *Prog. Polym. Sci.*, 28, 6, 963–1014, 2003.
5. Bian, H., Cao, M., Wen, H., Tan, Z., Jia, S., Cui, J., Biodegradation of polyvinyl alcohol using cross-linked enzyme aggregates of degrading enzymes from Bacillus niacini. *Int. J. Biol. Macromol.*, 124, 10–16, 2019.
6. Olsén, P., Herrera, N., Berglund, L.A., Toward biocomposites recycling: Localized interphase degradation in PCL-cellulose biocomposites and its mitigation. *Biomacromolecules*, 21, 5, 1795–1801, 2020.
7. Rathinavel, S. and Saravanakumar, S.S., Development and analysis of poly vinyl alcohol/orange peel powder biocomposite films. *J. Nat. Fibers*, 18, 1, 1–10, 2020.
8. Kurniawan, A., Muneekaew, S., Hung, C.-W., Chou, S.-H., Wang, M.-J., Modulated transdermal delivery of nonsteroidal anti-inflammatory drug by macroporous poly(vinyl alcohol)-graphene oxide nanocomposite films. *Int. J. Pharm.*, 566, 708–716, 2019.
9. Mallakpour, S. and Khadem, E., Facile and cost-effective preparation of PVA/modified calcium carbonate nanocomposites via ultrasonic irradiation: Application in adsorption of heavy metal and oxygen permeation property. *Ultrason. Sonochem.*, 39, 430–438, 2017.
10. Liu, X., Chen, X., Ren, J., Chang, M., He, B., Zhang, C., Effects of nano-ZnO and nano-SiO_2 particles on properties of PVA/xylan composite films. *Int. J. Biol. Macromol.*, 132, 978–986, 2019.

11. Zhang, W., Yang, X., Li, C., Liang, M., Lu, C., Deng, Y., Mechanochemical activation of cellulose and its thermoplastic polyvinyl alcohol ecocomposites with enhanced physicochemical properties. *Carbohydr. Polym.*, 83, 1, 257–263, 2011.
12. Lisboa, H., Reinforcement of poly(vinyl alcohol) films with alpha-chitin nanowhiskers. *Polímeros: Ciencia e Technologia*, 28, n1, 69–75, 2018.
13. Irvin, C.W., Satam, C.C., Carson Meredith, J., Shofner, M.L., Mechanical reinforcement and thermal properties of PVA tricomponent nanocomposites with chitin nanofibers and cellulose nanocrystals. *Compos. Part A: Appl. Sci. Manuf.*, 116, 147–157, 2019.
14. Perumal, A.B., Sellamuthu, P.S., Nambiar, R.B., Sadiku, E.R., Development of polyvinyl alcohol/chitosan bio-nanocomposite films reinforced with cellulose nanocrystals isolated from rice straw. *Appl. Surf. Sci.*, 449, 591–602, 2018.
15. Hajji, S., Chaker, A., Jridi, M., Maalej, H., Jellouli, K., Boufi, S., Nasri, M., Structural analysis, and antioxidant and antibacterial properties of chitosan-poly(vinyl alcohol) biodegradable films. *Environ. Sci. Pollut. Res.*, 23, 15, 15310–15320, 2016.
16. Kumar, A. and Ha, S.S., PVA-based hydrogels for tissue engineering: A review. *Int. J. Polym. Mater. Polym. Biomater.*, 66, 159–182, 2017.
17. W.O. Herrmann and W. Haehnel, U.S. Pat. 1,672,156, 1928.
18. Nangia, R., Shukla, N.K., Sharma, A., Preparation, structural and dielectric properties of solution grown polyvinyl alcohol (PVA) film. *IOP Conf. Ser.: Mater. Sci. Eng.*, 225, 7–15, 2017.
19. Hyon, S.H., Cha, W.I., Ikada, Y., Kita, M., Ogura, Y., Honda, Y., Poly(vinyl alcohol) hydrogels as soft contact lens material. *J. Biomater. Sci. Polym.*, 5, 397–406, 1994.
20. Jiang, H., Campbell, G., Boughner, D., Wan, W.K., Quantz, M., Design and manufacture of a polyvinyl alcohol (PVA) cryogel tri-leaflet heart valve prosthesis. *Med. Eng. Phys.*, 4, 269–277, 2004.
21. Kayal, S. and Ramanujan, R.V., Doxorubicin loaded PVA coated iron oxide nanoparticles for targeted drug delivery. *Mater. Sci. Eng.: C*, 30, 484–490, 2010.
22. Kokabia, M., Sirousazarb, M., Hassanc, Z.M., PVA–clay nanocomposite hydrogels for wound dressing. *Eur. Polym. J.*, 43, 773–781, 2007.
23. Mohanapriya, S. and Raj, V., Tuning biological properties of poly(vinyl alcohol) with amino acids and studying its influence on osteoblastic cell adhesion. *Mater. Sci. Eng. C*, 86, 70–82, 2018.

24. Mohanapriya, S., Mumjitha, M., Purnasai, K., Raj, V., Fabrication and characterization of poly(vinyl alcohol)-TiO_2 nanocomposite films for orthopaedic applications. *Mech. Behav. Biomed. Mater.*, 63, 141–156, 2016.
25. Vasanthakumar, V., Mohanapriya, S., Priyadharsan, A., Anbarasan, P.M., Nambissan, P.M.G., Raj, V., Tunable physicochemical and free volume characteristics of novel terpolymer–poly(vinylalcohol)-grafted membranes for direct methanol fuel cells. *New J. Chem.*, 43, 2942–2952, 2019.
26. Mohanapriya, S., Bhat, S.D., Sahu, A.K., Manokaran, A., Vijaykumar, R., Pitchumani, S., Sridhar, P., Shukla, A.K., Sodium-alginate-based proton-exchange membranes as electrolytes for DMFCs. *Energy Environ. Sci.*, 3, 1746–1756, 2010.
27. Mohanapriya, S., Rambabu, G., Suganthi, S., Bhat, S.D., Vasanthkumar, V., Anbarasu, V., Raj, V., Bio-functionalized hybrid nanocomposite membranes for direct methanol fuel cells. *RSC Adv.*, 6, 57709–57721, 2016.
28. Suganthi, S., Mohanapriya, S., Raj, V., Bio composite proton exchange membrane electrolytes for direct methanol fuel cells. *J. Appl. Polym. Sci.*, 133, 43514, 2016.
29. Mohanapriya, S., Rambabu, G., Bhat, S.D., Raj, V., Pectin based nanocomposite membranes as green electrolytes for direct methanol fuel cells. *Arabian J. Chem.*, 13, 2024–2040, 2020.
30. Mohanapriya, S., Rambabu, G., Bhat, S.D., Raj, V., Hybrid membranes for polymer electrolyte fuel cells operating under various relative humidity values. *J. Solid State Electrochem.*, 21, 3437–3448, 2017.
31. Abu, N., Yern, C., Chuah, L., Thevy, C., Azowa, N., Thermal and dynamic mechanical properties of grafted kenaf filled poly(vinyl chloride)/ethylene vinyl acetate composites. *J. Mater.*, 65, 204–211, 2015.
32. Tan, B.K., Ching, Y.C., Poh, S.C., Abdullah, L.C., A review of natural fiber reinforced poly(vinyl alcohol) based composites: Application and opportunity. *Polym.*, 7, 2205–2222, 2015.
33. Taylor, P., Yin, J., Fan, H., Zhou, J., Cellulose acetate/poly(vinyl alcohol) and cellulose acetate/crosslinked poly(vinyl alcohol) blend membranes: Preparation, characterization, and antifouling properties. *Desalin. Water Treat.*, 57, 23, 1–13, 37–41, 2015.
34. Zhou, K., Zhang, Q.G., Han, G.L., Zhu, A.M., Liu, Q.L., Pervaporation of water–ethanol and methanol–MTBE mixtures using poly(vinyl alcohol)/cellulose acetate blended membranes. *J. Membr. Sci.*, 448, 93–101, 2013.
35. Bernal-ballén, A., Kuritka, I., Saha, P., Preparation and characterization of a bioartificial polymeric material: Bilayer of cellulose acetate-PVA. *J. Polym. Sci. A*, 2016, 1–11, 2016.

36. Nouri, M. and Marjani, A., Preparation of cellulose acetate membrane coated by PVA/Fe_3O_4 nanocomposite thin film: An *in situ* procedure. *Colloid Polym. Sci.*, 296, 1213–1223, 2018.
37. Petersen, N. and Gatenholm, P., Bacterial cellulose-based materials and medical devices: Current state and perspectives. *Appl. Microbiol. Biotechnol.*, 91, 1277–1286, 2011.
38. Yano, H., Optically transparent composites reinforced with networks of bacterial nanofibers. *Adv. Mater.*, 17, 153–155, 2005.
39. Hu, Y. and Catchmark, J.M., Formation and characterization of spherelike bacterial cellulose particles produced by *Acetobacter xylinum* JCM 9730 Strain. *Biomacromolecules*, 11, 1727–1734, 2010.
40. Jeon, S., Yoo, Y., Park, J., Kim, H., Hyun, J., Electrical conductivity and optical transparency of bacterial cellulose based composite by static and agitated methods. *Curr. Appl. Phys.*, 14, 1621–1624, 2014.
41. Barud, H.G.O., Barud, S., Cavicchioli, M., Silva, T., Batista, O., Junior, D.O., Santos, D.M., Luis, A., Almeida, D.O., Celes, F., Matos, V., De Oliveira, C.I., Francielli, P., Oliveira, D., Andrade, R., Crispim, D., Ribeiro, S.J.L., Preparation and characterization of a bacterial cellulose/silk fibroin sponge scaffold for tissue regeneration. *Carbohydr. Polym.*, 128, 41–51, 2015.
42. Panaitescu, D.M., Frone, A.N., Chiulan, I., Nanostructured biocomposites from aliphatic polyesters and bacterial cellulose. *Ind. Crops Prod.*, 93, 251–266, 2016.
43. Lu, M., Guan, X.-H., Xu, X.-H., Wei, D.-Z., Characteristic and mechanism of Cr(VI) adsorption by ammonium sulfamate-bacterial cellulose in aqueous solutions. *Chin. Chem. Lett.*, 24, 253–256, 2013.
44. Abral, H., Hartono, A., Ha, F., Handayani, D., Sugiarti, E., Characterization of PVA/cassava starch biocomposites fabricated with and without sonication using bacterial cellulose fiber loadings. *Carbohydr. Polym.*, 206, 593–601, 2019.
45. Wong, S., Kasapis, S., Huang, D., Molecular weight and crystallinity alteration of cellulose via prolonged ultrasound fragmentation. *Food Hydrocolloids*, 26, 365–369, 2012.
46. Frone, A.N., Panaitescu, D.M., Donescu, D., Spataru, C.I., Radovici, C., Trusca, R., Somoghi, R., Preparation and characterization of PVA composites with cellulose nanofibers obtained by ultrasonication. *Bioresources*, 6, 487–512, 2011.
47. Id, M.A., Abral, H., Kasim, A., Pratoto, A., Mahardika, M., Mechanical properties of a water hyacinth nanofiber cellulose reinforced thermoplastic starch bionanocomposite: Effect of ultrasonic vibration during processing. *Fibers*, 6, 1–9, 2018.

48. Gea, S., Bilotti, E., Reynolds, C.T., Soykeabkeaw, N., Peijs, T., Bacterial cellulose – poly(vinyl alcohol) nanocomposites prepared by an *in-situ* process. *Mater. Lett.*, 64, 901–904, 2010.
49. Lu, Y., Liu, Y., Yu, H., Sun, Q., Preparation, characterization and biocompatibility of regenerated cellulose/PVA blend membranes in 1-allyl-3-methyimidazolium chloride. *Adv. Mater. Res.*, 180, 1108–1111, 2011.
50. Abdulkhani, A., Hojati, E., Ashori, A., Hamzeh, Y., Naghi, A., Preparation of cellulose/polyvinyl alcohol biocomposite films using 1-n-butyl-3-methylimidazolium chloride. *Int. J. Biol. Macromol.*, 62, 379–386, 2013.
51. Kaco, H., Zakaria, S., Chia, C.H., Zhang, L., Transparent and printable regenerated kenaf cellulose/PVA film. *Bioresources*, 9, 2167–2178, 2014.
52. Ioana, R., Bercea, M., Avadanei, M., Lisa, G., Biliuta, G., Coseri, S., Green route for the fabrication of self-healable hydrogels based on tricarboxy cellulose and poly(vinyl alcohol). *Int. J. Biol. Macromol.*, 123, 744–751, 2019.
53. Bai, H., Li, Z., Zhang, S., Wang, W., Dong, W., Interpenetrating polymer networks in polyvinyl alcohol/cellulose nanocrystals hydrogels to develop absorbent materials. *Carbohydr. Polym.*, 200, 468–476, 2018.
54. Tien, N., Chollakup, R., Smitthipong, W., Nimchua, T., Utilizing cellulose from sugarcane bagasse mixed with poly(vinyl alcohol) for tissue engineering scaffold fabrication. *Ind. Crops Prod.*, 100, 183–197, 2017.
55. Nakashima, T., Xu, C., Bin, Y., Matsuo, M., Morphology and mechanical properties of poly(vinyl alcohol) and starch blends prepared by gelation/crystallization from solutions. *Colloid Polym. Sci.*, 279, 7, 646–654, 2001.
56. Huang, X. and Netravali, A., Biodegradable green composites made using bamboo micro/nano fibrils and chemically modified soy protein resin. *Compos. Sci. Technol.*, 69, 7, 1009–1015, 2009.
57. Witt, U., Müller, R.J., Deckwer, W.D., Biodegradation of polyester copolymers containing aromatic compounds. *J. Macromol. Sci., Part A*, 32, 4, 851–856, 1995.
58. Ishigaki, T., Kawagoshi, Y., Ike, M., Fujita, M., Biodegradation of polyvinyl-alcohol-starch blend plastic film. *World J. Microbiol. Biotechnol.*, 15, 3, 321–327, 1999.
59. Azahari, N.A., Othman, N., Ismail, H., Biodegradation studies of polyvinyl alcohol/corn starch blend films in solid and solution media. *J. Phys. Sci.*, 22, 2, 15–31, 2011.
60. Chiellini, E., Corti, A., Solaro, R., Biodegradation of poly(vinyl alcohol) based blown films under different environmental conditions. *Polym. Degrad. Stab.*, 64, 305, 1999.
61. Faria, F.O., Vercelheze, A.E.S., Mali, S., Physical properties of biodegradable films based on cassava starch, polyvinyl alcohol and montmorillonite. *Quím. Nova*, 35, 3, 487–492, 2012.

62. Guohua, Z., Ya, L., Cuilan, F., Min, Z., Caiqiong, Z., Zongdao, C., Water resistance, mechanical properties and biodegradability of methylated-cornstarch/poly(vinyl alcohol) blend film. *Polym. Degrad. Stab.*, 91, 4, 703–711, 2006.
63. Taghizadeh, M.T. and Sabouri, N., Thermal degradation behavior of polyvinyl alcohol/starch/carboxymethyl cellulose/clay nanocomposites. *Univers. J. Chem.*, 1, 2, 21–29, 2013.
64. Zaaba, N.F., Ismail, H., Jaafar, M., The effects of modifying peanut shell powder with polyvinyl alcohol on the properties of recycled polypropylene and peanut shell powder composites. *BioResources*, 9, 2, 2128–2142, 2014.
65. Zou, G.-X., Jin, P.-Q., Xin, L.-Z., Extruded starch/PVA composites: Water resistance, thermal properties and morphology. *J. Elastomers Plast.*, 40, 4, 303–316, 2008.
66. Xiong, H., Tang, S., Zou, P., Tang, H., Effect of nano-SiO_2 on the performance of starch/polyvinyl alcohol blend films. *Carbohydr. Polym.*, 72, 3, 521–526, 2008.
67. Schmiele, M., Sampaio, U.M., Pedrosa Silva Clerici, M.T., Chapter 1 - Basic principles: Composition and properties of starch, in: *Starches for Food Application*, M.T.P. Silva Clerici and M. Schmiele (Eds.), pp. 1–22, Academic Press, USA, 2019.
68. Yun, Y.H., Wee, Y.J., Byun, H.S., Yoon, S.D., Biodegradability of chemically modified starch (RS4)/PVA blend films: Part 2. *J. Polym. Environ.*, 16, 1, 12–18, 2008.
69. Kondo, T., Sawatari, C., Manley, R.S.J., Gray, D.G., Characterization of hydrogen bonding in cellulose-synthetic polymer blends systems with regioselectively substituted methycellulose. *Macromol.*, 27, 1, 210–215, 1994.
70. Tang, S., Zou, P., Xiong, H., Tang, H., Effect of nano-SiO_2 on the performance of starch/polyvinyl alcohol blend films. *Carbohydr. Polym.*, 72, 3, 521–526, 2007.
71. Yun, Y.-H., Wee, Y.-W., Byun, H.-S., Yoon, S.-D., Biodegradability of chemically modified starch (RS4)/PVA blend films: Part 2. *J. Polym. Environ.*, 16, 1, 12–18, 2008.
72. Wang, H., Sun, X., Seib, P., Strengthening blends of poly(lactic acid) and starch with methylenediphenyl diisocyanate. *J. Appl. Polym. Sci.*, 82, 7, 1761–1767, 2001.
73. Kalambur, S. and Rizvi, S.S.H., An overview of starch-based plastic blends from reactive extrusion. *J. Plast. Film Sheeting*, 22, 1, 39–58, 2006.
74. Mani, R., Tang, J., Bhattacharya, M., Synthesis and characterization of starch-graft-polycaprolactone as compatibilizer for starch/polycaprolactone blends. *Macromol. Rapid Commun.*, 19, 283–286, 1998.

75. Halimatudahliana, Ismail, H., Nasir, M., The effect of various compatibilizers on mechanical properties of polystyrene/polypropylene blend. *Polym. Test.*, 21, 2, 163–170, 2002.
76. Mischnick, P. and Kuhn, G., Model studies methyl amyloses: Correlation between reaction conditions and primary structure. *Carbohydr. Res.*, 290, 199–207, 1996.
77. Liu, Z., Dong, Y., Men, H., Jiang, M., Tong, J., Zhou, J., Post-crosslinking modification of thermoplastic starch/PVA blend films by using sodium hexametaphosphate. *Carbohydr. Polym.*, 89, 473–477, 2012.
78. Sadhu, S.D., Anshuman Soni, S.G.V., Garg, M., Preparation of starch-polyvinyl alcohol (PVA) blend using potato and study of its mechanical properties. *Int. J. Pharm. Sci. Invent.*, 3, 3, 33–37, 2014.
79. Zhijun Wu, J.W., Peng, T., Li, Y., Lin, D., Xing, B., Chunxiao Li, Y.Y., Yang, L., Zhang, L., Ma, R., Wu, W., J.D., Lv, X., Han, G., Preparation and application of starch/polyvinyl alcohol/citric acid ternary blend antimicrobial functional food packaging films. *Polymers*, 9, 102, 1–19, 2017.
80. Ismail, H. and Zaaba, N.F., Effect of additives on properties of polyvinyl alcohol (PVA)/tapioca starch biodegradable films. *Polym. Plast. Technol. Eng.*, 50, 12, 1214–1219, 2011.
81. Tang, S., Peng Zou, H.X., Tang, H., Effect of nano-SiO_2 on the performance of starch/polyvinyl alcohol blend films. *Carbohydr. Polym.*, 72, 521–526, 2008.
82. Cataldo, F., On the ozone protection of polymers having non-conjugated unsaturation. *Polym. Degrad. Stab.*, 72, 287–296, 2001.
83. Rhim, J.W., Wang, L.F., Hong, S.I., Preparation and characterization of agar/AgNPs composite films with antimicrobial activity. *Food Hydrocoll.*, 33, 327–335, 2013.
84. DeMerlis, C.C. and Schoneker, D.R., Review of the oral toxicity of polyvinyl alcohol (PVA). *Food Chem. Toxicol.*, 41, 3, 319–326, 2003.
85. Liu, M., Guo, B., Du, M., Jia, D., Drying induced aggregation of halloysite nanotubes in polyvinyl alcohol/halloysite nanotubes solution and its effect on properties of composite film. *Appl. Phys. A*, 88, 2, 391–395, 2007.
86. Zhou, Y., Lei, L., Yang, B., Li, J., Ren, J., Preparation and characterization of polylactic acid (PLA) carbon nanotube nanocomposites. *Polym. Test.*, *68*, 34–38, 2018.
87. Hoidy, W.H., Ahmad, M.B., Al-Mulla, E.A.J., Ibrahim, N.A.B., Preparation and characterization of polylactic acid/polycaprolactone clay nanocomposites. *J. Appl. Sci.*, 10, 2, 97–106, 2010.
88. Cazón, P., Vázquez, M., Velazquez, G., Cellulose-glycerol-polyvinyl alcohol composite films for food packaging: Evaluation of water adsorption, mechanical properties, light-barrier properties and transparency. *Carbohydr. Polym.*, 195, 432–443, 2018.

89. Saini, P., Arora, M., Kumar, M.R., Poly(lactic acid) blends in biomedical applications. *Adv. Drug Delivery Rev.*, 107, 47–59, 2016.
90. Sun, W., Tian, J., Chen, L., He, S., Wang, J., Improvement of biodegradability of PVA-containing wastewater by ionizing radiation pretreatment. *Environ. Sci. Pollut. Res.*, 19, 8, 3178–3184, 2012.
91. Lin, C.C., Lee, L.T., Hsu, L.J., Degradation of polyvinyl alcohol in aqueous solutions using UV-365 nm/S2O82–process. *Int. J. Environ. Sci. Technol.*, 11, 3, 831–838, 2014.
92. Ye, B., Li, Y., Chen, Z., Wu, Q.Y., Wang, W.L., Wang, T., Hu, H.Y., Degradation of polyvinyl alcohol (PVA) by UV/chlorine oxidation: Radical roles, influencing factors, and degradation pathway. *Water Res.*, *124*, 381–387, 2017.
93. Rescignano, N., Fortunati, E., Montesano, S., Emiliani, C., Kenny, J.M., Martino, S., Armentano, I., PVA bio-nanocomposites: A new take-off using cellulose nanocrystals and PLGA nanoparticles. *Carbohydr. Polym.*, 99, 47–58, 2014.
94. Paradossi, G., Cavalieri, F., Chiessi, E., Spagnoli, C., Cowman, M.K., Poly(vinyl alcohol) as versatile biomaterial for potential biomedical applications. *J. Mater. Sci. Mater. Med.*, 14, 687–691, 2003.
95. Ramezanzadeh, B., Vakili, H., Amini, R., The effects of addition of poly(vinyl) alcohol (PVA) as a green corrosion inhibitor to the phosphate conversion coating on the anticorrosion and adhesion properties of the epoxy coating on the steel substrate. *Appl. Surf. Sci.*, 327, 174–181, 2015.
96. Hikku, G.S., Jeyasubramanian, K., Venugopal, A., Ghosh, R., Corrosion resistance behaviour of graphene/polyvinyl alcohol nanocomposite coating for aluminium-2219 alloy. *J. Alloys Compd.*, 716, 259–269, 2017.
97. DeMerlis, C. and Schoneker, D., Review of the oral toxicity of polyvinyl alcohol (PVA). *Food Chem. Toxicol.*, 41, 319–326, 2003.
98. Caló, E. and Khutoryanskiy, V.V., Biomedical applications of hydrogels: A review of patents and commercial products. *Eur. Polym. J.*, 65, 252–267, 2015.
99. Chee, B.S., De Lima, G.G., Devine, D., Nugent, M.J.D., Electrospun hydrogels composites for bone tissue engineering, in: *Application of Nanocomposite Materials in Orthopaedics*, p. 813740, 2018.
100. Uma Maheshwari, S., Samuel, V.K., Nagiah, N., Fabrication and evaluation of (PVA/HAp/PCL) bilayer composites as potential scaffolds for bone tissue regeneration application. *Ceram. Int.*, 40, 8469–8477, 2014.
101. Bao, C., Guo, Y., Song, L., Hu, Y., Poly(vinyl alcohol) nanocomposites based on graphene and graphite oxide: A comparative investigation of property and mechanism. *J. Mater. Chem.*, 21, 13942–13950, 2011.

102. Lin, C., Liu, Y., Xie, X., GO/PVA nanocomposites with significantly enhanced mechanical properties through metal ion coordination. *Chin. Chem. Lett.*, 30, 5, 1100–1104, 2019.
103. M.F., E.M., B.R., Fabrication and characterization of a starch-based nanocomposite scaffold with highly porous and gradient structure for bone tissue engineering. *Biomed. Phys. Eng. Express*, 4, 1–25, 2018.
104. Chakraborty, S., Ponrasu, T., Chandel, S., Dixit, M., Muthuvijayan, V., Reduced graphene oxide-loaded nanocomposite scaffolds for enhancing angiogenesis in tissue engineering applications. *R. Soc. Open Sci.*, 5, 1–14, 2018.
105. Kaur, T. and Thirugnanam, A., Tailoring *in vitro* biological and mechanical properties of polyvinyl alcohol reinforced with threshold carbon nanotube concentration for improved cellular response. *RSC Adv.*, 6, 39982–39992, 2016.
106. Millon, L.E., Oates, C.J., Wan, W., Compression properties of polyvinyl alcohol-bacterial cellulose nanocomposite. *J. Biomed. Mater. Res. - Part B Appl. Biomater.*, 90 B, 922–929, 2009.
107. Shamekhi, M.A., Mirzadeh, H., Mahdavi, H., Rabiee, A., Mohebbi-Kalhori, D., Baghaban Eslaminejad, M., Graphene oxide containing chitosan scaffolds for cartilage tissue engineering. *Int. J. Biol. Macromol.*, 127, 396–405, 2019.
108. Kanimozhi, K., Khaleel Basha, S., Sugantha Kumari, V., Kaviyarasu, K., Maaza, M., *In vitro* cytocompatibility of chitosan/PVA/methylcellulose – Nanocellulose nanocomposites scaffolds using L929 fibroblast cells. *Appl. Surf. Sci.*, 449, 574–583, 2018.
109. Mc Gann, M.J., Higginbotham, C.L., Geever, L.M., Nugent, M.J.D., The synthesis of novel pH-sensitive poly(vinyl alcohol) composite hydrogels using a freeze/thaw process for biomedical applications. *Int. J. Pharm.*, 372, 154–61, 2009.
110. Nugent, M.J.D. and Higginbotham, C.L., Preparation of a novel freeze thawed poly(vinyl alcohol) composite hydrogel for drug delivery applications. *Eur. J. Pharm. Biopharm.*, 67, 377–386, 2007.
111. Nugent, M.J.D., Hanley, A., Tomkins, P.T., Higginbotham, C.L., Investigation of a novel freeze-thaw process for the production of drug delivery hydrogels. *J. Mater. Sci. Mater. Med.*, 16, 1149–1158, 2005.
112. Kalita, H., Pal, P., Dhara, S., Pathak, A., Fabrication and characterization of polyvinyl alcohol/metal (Ca, Mg, Ti) doped zirconium phosphate nanocomposite films for scaffold-guided tissue engineering application. *Mater. Sci. Eng. C*, 71, 363–371, 2017.
113. Gonzalez, J.S., Maiolo, A.S., Hoppe, C.E., Alvarez, V.A., Composite gels based on Poly(vinyl alcohol) for biomedical uses. *Proc. Mater. Sci.*, 1, 483–490, 2012.

114. Hajeeassa, K.S., Hussein, M.A., Anwar, Y., Tashkandi, N.Y., Al-amshany, Z.M., Nanocomposites containing polyvinyl alcohol and reinforced carbon-based nanofiller: A super effective biologically active material. *Nanobiomedicine,* 5, 1–12, 2018.
115. Gautam, A. and Ram, S., Preparation and thermomechanical properties of Ag-PVA nanocomposite films. *Mater. Chem. Phys.*, 119, 266–271, 2010.
116. Jarosz, M., Kapusta-Kołodziej, J., Pawlik, A., Syrek, K., Sulka, G.D., Drug delivery systems based on titania nanostructures, in: *Nanostructures for Drug Delivery, Micro and Nano Technologies*, pp. 299–236, Elsevier Inc., 2017.
117. Mei, L., Zhang, Z., Zhao, L., Huang, L., Yang, X.L., Tang, J., Feng, S.S., Pharmaceutical nanotechnology for oral delivery of anticancer drugs. *Adv. Drug Deliv. Rev.*, 65, 880–890, 2013.
118. Tiwari, G., Tiwari, R., Sriwastawa, B., Bhati, L., Pandey, S., Pandey, P., Bannerjee, S.K., Drug delivery systems: An updated review. *Int. J. Pharm. Investig.*, 2, 2–11, 2012.
119. Mu, S., Li, G., Liang, Y., Wu, T., Ma, D., Hyperbranched polyglycerol-modified graphene oxide as an efficient drug carrier with good biocompatibility. *Mater. Sci. Eng. C*, 78, 639–646, 2017.
120. Bai, H., Li, C., Wang, X., Shi, G., A pH-sensitive graphene oxide composite hydrogel. *Chem. Commun.*, 46, 2376–2378, 2010.
121. Mahmoudi, M., Sant, S., Wang, B., Laurent, S., Sen, T., Superparamagnetic iron oxide nanoparticles (SPIONs): Development, surface modification and applications in chemotherapy. *Adv. Drug Deliv. Rev.*, 63, 24–46, 2011.
122. Arruebo, M., Fernández-pacheco, R., Ibarra, M.R., Santamaría, J., *Nanotoday,* 2, 22–32, 2007.
123. Ishtiaq, S., Ullah, A., Ali, K., Attaullah, M., Khan, H., Ali, H., Amjad, M., Tahir, M., Javed, M., Ghramh, H.A., Adgaba, N., Kanta, C., Composition and functional properties of propolis (bee glue): A review. *Saudi J. Biol. Sci.*, 26, 7, 1695–1703, 2018.
124. Bhowmick, S., Mohanty, S., Koul, V., Fabrication of transparent quaternized PVA/silver nanocomposite hydrogel and its evaluation as an antimicrobial patch for wound care systems. *J. Mater. Sci. Mater. Med.*, 27, 11, 160, 2016.
125. Rafienia, M., Saberi, A., Poorazizi, E., A novel fabrication of PVA/alginate-bioglass electrospun for biomedical engineering application. *Nanomed. J.*, 4, 152–163, 2017.
126. Tarun, K. and Gobi, N., Calcium alginate/PVA blended nano fibre matrix for wound dressing. *Indian J. Fibre Text. Res. (IJFTR),* 37, 127–132, 2012.
127. Noori, S., Kokabi, M., Hassan, Z.M., Nanoclay enhanced the mechanical properties of poly(vinyl alcohol)/chitosan/montmorillonite nanocomposite hydrogel as wound dressing. *Proc. Mater. Sci.*, 11, 152–156, 2015.

128. Kokabi, M., Sirousazar, M., Hassan, Z.M., PVA-clay nanocomposite hydrogels for wound dressing. *Eur. Polym. J.*, 43, 773–781, 2007.
129. Ray, S.S. and Okamoto, M., Polymer/layered silicate nanocomposites: A review from preparation to processing. *Prog. Polym. Sci.*, 28, 1539–1641, 2003.
130. Ali, M.H.M., Kahder, M.M., Al-saad, K.A., Al-meer, S., Properties of nanoclay PVA composites materials. *Science Connect*, 1, 1–9, 2013.
131. Koosha, M., Mirzadeh, H., Shokrgozar, M.A., Farokhi, M., Nanoclay-reinforced electrospun chitosan/PVA nanocomposite nanofibers for biomedical applications. *RSC Adv.*, 5, 10479–10487, 2015.
132. Bao, L., Hong, F.F., Chen, L., Li, X., Tang, J., Potential of PVA-doped bacterial nano-cellulose tubular composites for artificial blood vessels. *J. Mater. Chem. B*, 3, 8537–8547, 2015.
133. Mohanapriya, S., Mumjitha, M., PurnaSai, K., Raj, V., Fabrication and characterization of poly(vinyl alcohol)-TiO_2 nanocomposite films for orthopedic applications. *J. Mech. Behav. Biomed. Mater.*, 63, 141–156, 2016.
134. Parsa, P., Paydayesh, A., Davachi, S.M., Investigating the effect of tetracycline addition on nanocomposite hydrogels based on polyvinyl alcohol and chitosan nanoparticles for specific medical applications. *Int. J. Biol. Macromol.*, 121, 1061–1069, 2019.
135. Enayati, M.S., Behzad, T., Sajkiewicz, P., Rafienia, M., Bagheri, R., Ghasemi-Mobarakeh, L., Kolbuk, D., Pahlevanneshan, Z., Bonakdar, S.H., Development of electrospun poly(vinyl alcohol)-based bionanocomposite scaffolds for bone tissue engineering. *J. Biomed. Mater. Res. Part A*, 106, 1111–1120, 2018.
136. Koosha, M., Raoufi, M., Moravvej, H., One-pot reactive electrospinning of chitosan/PVA hydrogel nanofibers reinforced by halloysite nanotubes with enhanced fibroblast cell attachment for skin tissue regeneration. *Colloids Surf. B Biointerfaces*, 179, 270–279, 2019.
137. Tamburaci, S., Cecen, B., Ustun, O., Ergur, B.U., Havitcioglu, H., Tihminlioglu, F., Production and characterization of a novel bilayer nanocomposite scaffold composed of chitosan/Si-nHap and Zein/POSS structures for osteochondral tissue regeneration. *ACS Appl. Bio Mater.*, 2, 4, 1440–1455, 2019.
138. Prosanov, I.Y., Bulina, N.V., Gerasimov, K.B., Complexes of polyvinyl alcohol with insoluble inorganic compounds. *Phys. Solid State*, 55, 2132–35, 2013.
139. Yokoi, H., Kawata, S., Iwaizumi, M., Interaction modes between heavy metal ions and water-soluble polymers. 1. Spectroscopic and magnetic reexamination of the aqueous solutions of cupric ions and poly(vinyl alcohol). *J. Am. Chem. Soc.*, 108, 3358–3361, 1986.
140. Takahama, T., Saharin, S.M., Tashiro, K., Details of the intermolecular interactions in poly(vinyl alcohol)-iodine complexes as studied by quantum chemical calculations. *Polymer*, 99, 566–579, 2016.

141. Amo-Ochoa, P. and Zamora, F., Coordination polymers with nucleobases: From structural aspects to potential applications. *Coord. Chem. Rev.*, 276, 34–58, 2014.
142. Pettinari, C., Tabacaru, A., Galli, S., Coordination polymers and metal–organic frameworks based on poly(pyrazole)-containing ligands. *Coord. Chem. Rev.*, 307, 1–31, 2016.
143. Hassan, J.J., Mahdi, M.A., Chin, C.W., Hassan, Z., Abu-Hassan, H., Microwave assisted chemical bathdeposition of verticallyaligned ZnO nanorods on a variety of substrates seeded by PVA–$Zn(OH)_2$ nanocomposites. *Appl. Surf. Sci.*, 258, 4467–4472, 2012.
144. Wang, M., Lian, Y., Wang, X., PPV/PVA/ZnO nanocomposite prepared by complex precursor method and its photovoltaic application. *Curr. Appl. Phys.*, 9, 189–194, 2009.
145. Wang, H., Lu, X., Zhao, Y., Wang. Preparation, C., and characterization of ZnS: Cu/PVA composite nanofibers via electrospinning. *Mater. Lett.*, 60, 2480–2484, 2006.
146. Yu Prosanov, I., Benassi, E., Bulina, N.V., Structure of hybrid interpolymeric complexes of zinc halides with polyvinyl alcohol. *Curr. Appl. Polym. Sci.*, 2, 2018.
147. Haiyan, M., Yuguang, L., Hong, Y., Qiang, L., Minghao, R., Xubi, G., Ying, G., Yinfeng, L., Determination of coordination parameters of Cd^{2+} and Zn^{2+} with polyvinyl alcohol and chitosan. *J. Macromol. Sci., Part B: Phys.*, 50, 1, 132–143, 2010.
148. Yu, P.I. and Bulina, N.V., Polymeric sulfides CdS, CuS, and NiS in polyvinyl alcohol matrix. *Phys. Solid State*, 56, 1270–1272, 2014.
149. Yu, P.I., Abdulrahman, S.T., Thomas, S., Bulina, N.V., Gerasimov, K.B., Complex of polyvinyl alcohol with boric acid: Structure and use. *Mater. Today Commun.*, 14, 77–81, 2018.
150. Yu Prosanov, I., Benassi, E., Bulina, N.V., Matvienko, A.A., Gerasimov, K.B., Sidelnikov, A.A., Polymeric copper oxide: Preparation and investigation of its structure and optical properties. *J. Inorg. Organomet. Polym. Mater.*, 28, 2328–2335, 2018.
151. Yu, P.I., Complex compound polyvinyl alcohol-titanic acid/titanium dioxide. *Phys. Solid State*, 55, 450–453, 2013.
152. Yu, P.I., Yu Chesalov, A., Gerasimov, K.B., Bulina, N.V., Hybrid material polyvinyl alcohol-stannic acid/stannic oxide. *Phys. Solid State*, 55, 1519–1524, 2013.

2

Biodegradation Study of Polyvinyl Alcohol-Based Biocomposites and Bionanocomposites

Zahid Majeed[1]*, Muhammad Mubashir[2], Pau Loke Show[3] and Eefa Manzoor[1]

[1]Department of Biotechnology, University of Azad Jammu and Kashmir, Chellah Campus, Muzaffarabad, Azad Kashmir, Pakistan
[2]Department of Chemical Engineering and Chemistry, Eindhoven University of Technology, Eindhoven, Netherlands
[3]Department of Chemical and Environmental Engineering, University of Nottingham Malaysia, Semenyih, Selangor Darul Ehsan, Malaysia

Abstract

Increment in the world population has threatened the environment because of pollutants especially that which comes from nonbiodegradable synthetic toxic and polymers. Therefore, the denial of importance in industrial production of products of daily life mad thereof of these products efforts is concentrated to find alternate means to replace with green biomass-based biocomposites. Biocomposites possess significant advantages including cheap, less toxic, biodegradability and valuable functional properties, which make them attractive candidates for agriculture, health, energy generation, food and environmental applications. The growing concerns for environmental safety have directed to the prompt manufacturing of biobased biocomposites which are amenable to biodegrade. Optimization of properties with desired design has set a practical arduous task for advances in bio/nanocomposite science. Unique morphology, bioactive and stimuli-responsive biocomposites are manufactured through an appropriate selection of nanofillers, crosslinkers, plasticizers, additives to improve tensile strength, water absorbance, biocompatibility, porous/non-porous surface area, biochemical selectivity and reactivity, hydrophobicity/hydrophilicity. These important attractive properties are applied as criteria to assess the resistance of biocomposites against microorganisms and their enzymes. In this chapter, the

**Corresponding author*: zahidfdb@gmail.com; zahid.majeed@ajku.edu.pk

Visakh P. M. and Olga B. Nazarenko (eds.) Polyvinyl Alcohol-Based Biocomposites and Bionanocomposites, (31–58) © 2023 Scrivener Publishing LLC

contemporary review presents advances in the biodegradability of PVA and its bio/nanocomposites with cellulose, chitin, starch, hemicellulose, polylactic acid, and polyhydroxyalkanoate. Overall, it has been concluded that biodegradation analysis of the biobased composites/nanocomposites is critical for predicting performance under complex natural environments. Unfortunately, biodegradation is more often reported through testing weight loss. It is recommended that this method has to be replaced with a more comprehensive, well-designed assay, which includes pH, temperature, microbial/enzyme diversity, and contact time to predict accurate reaction rates for different stages of biodegradation under diverse soil or aquatic environment.

***Keywords*:** Biocomposites, bionanocomposites, biodegradation, polyvinyl alcohol, cellulose, chitin, starch, hemicellulose, polylactic acid, polyhydroxyalkanoates

2.1 Introduction

Industrial progress is directly linked to devastation of the environment because of the common use of synthetic plastic products in daily use. Plastic usage involves various risks, which are water contamination, food and animal's toxicity. Polyvinyl alcohol (PVA) ends up in water reservoirs, where it is responsible for the chronic toxicity effect on fauna and flora, which make up part of the aquatic ecosystem. Thus, it is required to replace plastic-based on nondegradable materials with alternates derived from green processes in order to diminish the deterioration of the environment by minimizing the toxicity after dumping as waste. PVA has been known for various applications in the industry with the concern about its environmental pollution. This has imposed a serious challenge and needs the attention of scientist to find novel methods for the biodegradation of PVA. PVA has properties such as thermal stability, water solubility, film forming, high viscosity, emulsifying, tensile strength, and flexibility [1]. PVA applications cover the research areas of formulation films, synthesis of coatings, and adhesives products, emulsion polymerization [1]. Globally, PVA production and consumption were assessed to be nearly 1.124 million tons in 2016.

As shown in Figure 2.1, major routes for biodegradation of any polymer pass through biodeterioration, depolymerization, bioassimilation, and mineralization [2]. In biodeterioration, the microbial biofilm populates

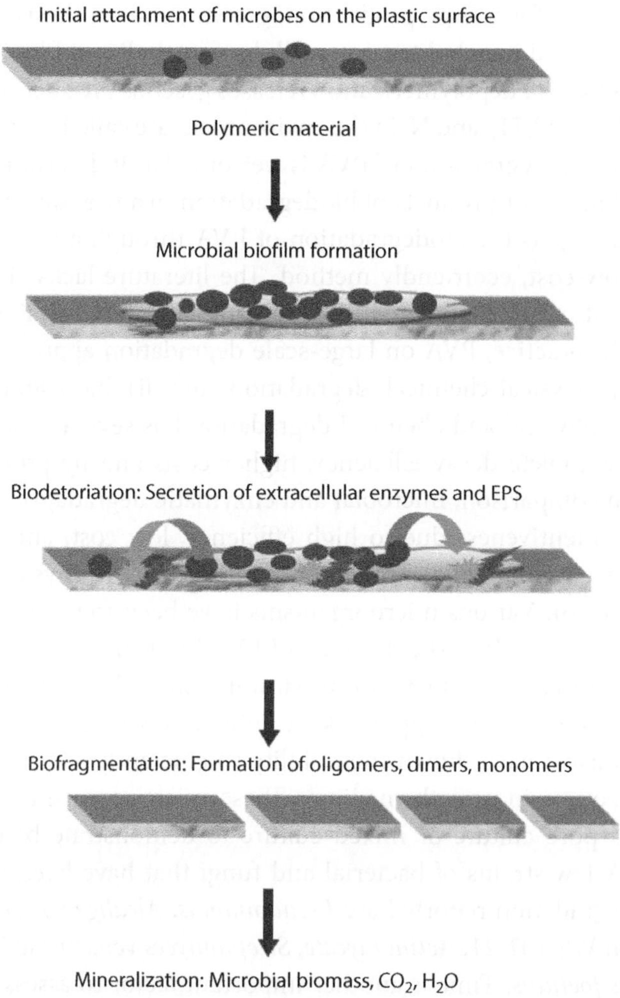

Figure 2.1 Scheme of biodegradation routes taken by microorganisms (Ganesh Kumar *et al.*, 2020 [3]), Reprinted after permission was granted from Elsevier.

the surface of the substrate on which they initiate apparent biodegradation. This changes the morphology of the surface into more rough and deformed. Later, depolymerization involved extracellular enzymes that are secreted by microbial cells. Enzymes catalyze the breakdown of bonds in polymers and produce low molecular weight products of like oligomers, dimers or monomers. Bioassimilation is the active uptake of low molecular weight depolymerization products into the microbial cell. These products are metabolized for energy production, growth, and synthesis of new cellular primary and metabolites. Intracellular metabolism of low molecular weight products of depolymerization releases gaseous products in the form of CO_2, CH_4, H_2O, H_2, and N_2 in the environment are called mineralization.

Microbial biodegradation of PVA is welcomed as it determines the safe removal of its major products of biodegradation in a friendly manner. The major challenge is the biodegradation of PVA through a more appropriate, safe, low-cost, ecofriendly method. The literature lacks effective and pertinent natural or engineered PVA-degrading bacterial strains and their enzymes. In practice, PVA on large-scale degradation approaches chiefly include, i) physical-chemical degradation and ii) biodegradation [4]. Meanwhile physical and chemical degradation has several disadvantages, such as incomplete decay efficiency, higher costs and by-product pollution [5]. In comparison, microbial and enzymatic degradation is drawing increasing attentiveness due to high efficiency, low cost, and more economic and environmental protection compared with physical-chemical degradation [6]. Various microorganisms have been found useful for the biodegradation of PVA. The diversity of PVA biodegraders has been cited in the literature, which spans from natural sources like activated sludge, soil, and biodegradable support like a polymeric sheets. These strains of microorganisms have shown their ability to efficiently assimilate PVA as a carbon source in growth medium. These microorganisms are studied either as a pure culture or mixed culture to demonstrate better activity on PVA. A few strains of bacterial and fungi that have been utilized for PVA biodegradation reported are *Pseudomonas*, *Alcaligenes*, and *Bacillus*. *Penicillium* WSH02-21, *Actinomycete*, *Streptomyces venezuelae* GY1 strain. *Aspergillus foetidus*. Time is another important factor to assess the rate of biodegradation, which may be as short as 12 days up to 15 weeks (a consortium of PVA-degrading *Vibrio* sp.) [7].

The biodegradation of PVA-based materials has been thoroughly discussed in earlier advanced work on the biochemical reactions and microbial growth dynamics for PVA blends, composites, and copolymers biodegradation and influence environmental conditions. PVA biodegradation proceeds in a random chain cleavage mode in which enzymes oxidize the carbon-carbon linkages in the PVA [8].

The two enzymes that participate in PVA's cleavage have been revealed which are oxidases and/or hydrolases [9]. A two-step route for the biodegradation of PVA has been suggested [1, 9].

First, oxidized hydroxyl groups change into diketone/monoketone moieties, by the activity of secondary alcohol oxidases (SAO) in addition to dehydrogenase. The second hydrolysis of the carbonyl structure takes places with oxidized PVA hydrolase (OPH). Unusual α/β-hydrolase structures of OPH have been studied in two different known PVA-degrading organisms, which belonged to the bacterial species *Sphingopyxis* sp. 113P3 and *Pseudomonas* sp. VM15C. The active site of the enzyme is known to be enclosed by a lid-like β-ribbon. OPH targets cleavage of the C-C bond present in β-diketone possibly via a double-oxyanion-hole mechanism [10]. Bioinformatics studies on the strain of bacteria *Stenotrophomonas rhizophila* QL-P4S have identified genes that encode PVA/vinyl alcohol oligomer (OVA)-degrading enzymes. Such enzymes have high catalytic activity and possess the ability to biodegrade both PVA and OVA due to oxidase and exocrine activity. The expression of the PVA enzymes can be inducible under appropriate conditions. Bacterial species that utilize PVA have been found in sludge samples by providing PVA is a selective source of carbon. PVAase, an enzyme that degrades PVA, secreted from *Bacillus niacin* immobilized by cross-linking as enzyme aggregates have shown improved enzyme activity by approximately 90% compared with free PVAase. Unpurified PVAase can increase the usability of its large-scale application in the industry [11]. Table 2.1 illustrates the succinct view of developments in the biodegradation of PVA alcohol and its biocomposites with their specific strains, enzyme, and experimental conditions.

Bacterial mode of PVA biodegradation comprises intracellular and extracellular mechanism. In case of intracellular degradation, oxidation of PVA is started by PVA dehydrogenase (PVADHs) with cytochrome c as an electron acceptor. OPH can participate in hydrolysis of the diketone of oxiPVA. Diketone is converted into a methyl ketone and aldehyde products. OVA (vinyl alcohol oligomer) is generated from PVA after degradation intracellularly by the action of secondary alcohol dehydrogenases (SADHs) [12].

The low molecular weight form of PVA is uptaken and has undergone the assimilation in bacterial cell's periplasm, which is laden with secondary alcohol dehydrogenases (SADHs). A low molecular weight form of PVA is uptaken and has undergone the assimilation in bacterial cell's periplasm, which is laden with key oxidizing enzymes as shown in Figure 2.2. These oxidizing enzymes are i) pyrroloquinoline quinone (PQQ)-dependent PVA

Table 2.1 Microbial biodegradation of PVA alcohol and its biocomposites with their experimental conditions.

PVA/biocomposites	Microorganisms	Enzyme	Biodegradation assay	Source	Ref.
PVA	*Bacillus niacini*	PVAase	Immobilization	Activated sludge	Bian *et al.* (2019) [11]
PVA	*Stenotrophomonas rhizophila* QL-P4	PVA oxidase	Bioinformatics	Fallen leaves	Wei *et al.* (2018) [12]
Poly(vinyl alcohol) with starch/ montmorillonites	-	α-Amylase	Liquid medium	Commercial (Sigma Aldrich)	Spiridon *et al.* (2008) [46]
PVA and Bacterial Cellulose Composites	*Aspergillus niger 105*	-	Liquid medium	Microbial Collection of INCDCP-ICECHIM (Romania)	Stoica-Guzun *et al.* (2011) [76]
PVA	*Janthinobacterium* sp. (WSH04-01)	Crude PVAase	Liquid medium	Soil samples	Du *et al.* (2007) [6]
PVA-graft-lactic acid copolymers	*Trichotecium roseum*	-	zone of clearance	Plant residues	Lipsa *et al.* (2015) [77]

(Continued)

Table 2.1 Microbial biodegradation of PVA alcohol and its biocomposites with their experimental conditions. (*Continued*)

PVA/biocomposites	Microorganisms	Enzyme	Biodegradation assay	Source	Ref.
PVA low-linear density polyethylene-blended	*Vibrio alginolyticus and Vibrio*	DNase, lipase, gelatinase, caseinase and amylase, pectinase, cellulases	zone of clearance, tensile strength, Scanning electron micrograph	Marine Sediment	(Raghul *et al.*, 2014) [7]
chitosan/polyvinyl alcohol/cellulose nanocomposite	-	-	Weight loss	-	(Samzadeh-Kermani and Esfandiary, 2016) [24]
Maize Starch-PVA/Clay Nanocomposite films	-	-	Gravimetrical tests and macroscopic observations		(Katerinopoulou *et al.*, 2019) [41]
PVA starch-cellulosic composite	-	-	Soil burial test		(Mittal, Garg, and Bajpai, 2020) [39]

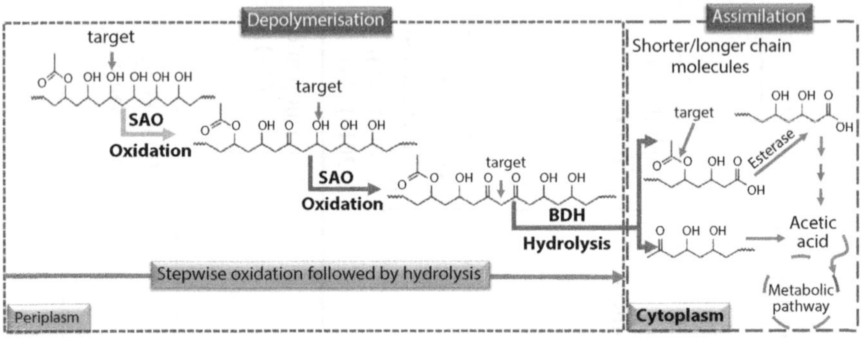

Figure 2.2 Intracellular reactions of PVA depolymerization and assimilation in bacteria.

dehydrogenase, ii) OPH [13], and iii) diketone hydrolase (BDH). During oxidation, PVA is attacked by intracellular pyrroloquinoline quinone (PQQ)-dependent PVADH. PVADH activity requires cytochrome c as an electron acceptor. Further hydrolysis process is taken over by OPH or BDH [14]. Extracellular oxidation of PVA involves two steps. In the first step, PVA oxidation requires the secretion of SAO with the help of O2 molecule as a strong electron acceptor. In the second phase, secreted BDH hydrolyzes the oxidized PVA. The PVA degradation needs an excess of PVA-degrading enzymes, which could not be met as an expression of PVADH reside in inclusion bodies. As a consequence, an insignificant amount is present in the extract or secreted right into the culture medium [15]. The exhaustive mechanistic scrutiny of PVA biodegradation is impeded because of the technical limitations of genome sequencing of specific strains.

2.2 Biodegradable PVA Biocomposites and Bionanocomposites

Composite is referred to as any multiphase composition of a material that is well-characterized and exhibits a better combination of properties that are contributed by both constituent phases. Moreover, the constituent phases must be chemically different and separated by a discrete interface [16].

Recyclable biocomposites derived through naturally degradable polymers are prepared to make composition attractive. This strategy enhances biocomposites properties after blending with another nano-size biodegradable filler materials for better processability, usability, and extending the life of application [17]. The Orange peel powder improved the properties of PVA films and make the PVA suitable for packaging application [18]. The biodegradability feature of PVA nanocomposite films lessened to a certain degree by modification with expensive inorganic nanoparticles of graphene oxide nanosheet [19] calcium carbonate nanoparticles [20], ZnO, and nano-SiO_2 [21]. These nanocomposite films have showned significant improvements in the barrier performance due to the presence of nanofillers. The search for renewable green biomass as a replacement for nanofiller is considered as an attractive solution for the biomanufacturing of fully biodegradable bionanocomposites. Biodegradability explains the post-application safe disposal in the environmental. Different biocomposites along with bionanocomposites of PVA with different biomass is presented in the following subsections.

2.2.1 PVA/Cellulose-Based Biocomposites and Bionanocomposites

Cellulose is a carbohydrate and the length of cellulose polymer depends on the source and variations occurs in a number of repeats which form the chain. Cellulose is not soluble in water due to strong inter- and intramolecular hydrogen bonding, which stabilizes the partially crystalline structure [22]. Cellulose is a renewable natural polymer that demonstrates attractive characteristics such as low cost, high biocompatibility, thermal recyclability by combustion, and good mechanical and biodegradability. On the other hand, *Aspergillus niger* (*A. niger*) strain is known to attack composite materials obtained from PVA and bacterial cellulose (BC). A composite film of PVA-BC with thickness of 0.02 mm, prepared by casting technique biodegrades with *A. niger* 105. Supplementation of the culture medium with easily accessible carbon source stimulates the *A. niger* 105 growth in biodegradation assay. The degree of microbial degradation is governed by the specifications of the culture medium and composition of PVA-BC. The addition of glucose in the culture medium and cellulose presence in PVA-BC enhances the growth and speeds up the biodegradation of PVA-BC.

In another study, the mechanochemical reactivity of cellulose and thermoplastic made of PVA ecocomposites with better physicochemical properties investigated for biodegradation. In PVA/cellulose ecocomposites, the surface of mechanochemically activated cellulose has hydroxyl groups on cellulose surface benefit the interfacial contact. Biodegradability of PVA/

cellulose composites functioned at 22°C to 27°C and relative humidity ranges 70% to 80%. Samples displayed a fast weight loss in 16 days in a soil burial test. Weight loss has decelerated in the succeeding soil burial period after 16 days. Their work concludes that the cellulose biodegradability rates are higher than PVA in ecocomposites, which resulted in higher weight loss with better biodegradability than that of neat PVA. In biodegraded biocomposites, recovery and analysis of constituents make the composites from complex matrix like soil compromised the study outcome. For example, cellulose fibers collected from the soil after the first 16 days of biodegradation of PVA/cellulose in buried soil severely hinders the degree of biodegradation under natural decomposing conditions. PVA mixed with cellulose prepared through 40 cycles of pan-milling was more possible to biodegrade than cellulose obtained through a single cycle of pan milling. A higher number of cycles of pan-milling reduces the size of cellulose fibers <20 μm and is less crystalline, which promotes the substrates utilization activity for microorganisms through enhanced substrate-microorganism interaction in the soil [23]. Monocrystalline cellulose was used to synthesize biodegradable polar chitosan/polyvinyl alcohol/nanocrystalline cellulose (Cts/PVA/NCC) bio nanocomposites having a size of 85 nm [24]. The Cts/PVA/NCC nanocomposites loaded with concentrations of NCC 5%, 10%, and 15% were evaluated under compost state for weight loss for 5 days. NCC promotes the biodegradation susceptibility of Cts/PVA nanocomposites occurs at a higher rate in the natural environment. In another work, plasma treatment rendered the composites cellulose/PVA biocomposites more hydrophilic and biodegradable. Hydrophilicity allows the seepage of water and microorganism into biocomposites, which promote faster biodegradation [25].

2.2.2 PVA/Chitin-Based Biocomposites and Bionanocomposites

Chitin is a natural biopolymer and its role is to provide structural support to body parts in the living organisms that belong to fungi, insects, crustaceans, and algae [26]. The composition of chitin reveals the presence of (1→4)-β-linked *N*-acetyl-D-glucosamine (GlcNAc) bonding in its polymer backbone. Sugar units are rotated to each other at 180°. N,N′-diacetylchitobiose [$(GlcNAc)_2$], is a disaccharide that forms the structural unit in chitin. Chitosan (CS) is an equivalent to chitin. However, chitosan is deacetylated form of chitin with much of variation in its degree of deacetylation [27]. Degradation of chitin requires the hydrolysis of the (1→4)-β-glycoside bond by chitinase, an enzyme that catalyzes the degradation of chitin by a process known as chitinolytic. *In-situ* methods have been used to develop silica-reinforced PVA/liquefied chitin biocomposite as biodegradable films for its potential applications in food packaging. The silica reinforcement of polyvinyl alcohol/liquefied chitin blend films slightly decreased the biodegradation rate in the soil. Biodegradability is maintained at a significant level of 43% over the period of 30 days of soil burial.

In another work, a spraying technique for *in-situ* obtaining the PVA/chitosan-loaded silver nanocomposite hydrogel has been used for anti-bacterial applications. Such emergence of bionanocomposites hydrogel could show a suitable biodegradable dressing for applications in wound healing [28]. The composition of PVA/nanowhiskers biocomposites discloses its biodegradable and environmentaly friendly biomaterials with utilization as a food source after disposal in a manner which have a positive effect during natural degradability. Strengthening of PVA nanocomposites after the addition of alpha-chitin nanowhiskers has shown improved mechanical properties, which possibly increased the resistance to biodegradation [29]. Machine-driven reinforcement of PVA tricomponent nanocomposites with chitin nanofibers and cellulose nanocrystals are known for better thermal properties, which reduce the biodegradability under ambient temperate conditions [30]. In fact, tricomponent nanocomposites change their mechanical and thermal characteristics, which turn them into high-performance biomaterials with possible low environmental impact. Chitin nanofibers provide potential nanofiller of PVA for better knowledge-driven modulation of material features. In other scientific work, chitosan/polyvinyl alcohol/thiabendazoluim-montmorillonite bio-nanocomposite for active packing applications have shown excellent antimicrobial properties [31].

The antimicrobial test proved that these bionanocomposites have a microbial inhibitory effect on bacteria *Pseudomonas aeruginosa*, *Staphylococcus aureus*, and *Escherichia coli* compared to CS/PVA film. Based on the reported literature, PVA/CS blends have shown the plasticization and compatibilization between PVA and CS by use of glycerol/polyethylene dramatically reduced the antibacterial activity of blends [32]. This would make more suitable these blends for biodegradation by attracting microorganisms with less effect on antibacterial activity. This result could enhance the structural and biological properties of the CS/PVA nanocomposite film, which mimic appropriately the human bone extracellular matrix for application in human bone regeneration [33].

Environmentally useful biocomposites films are composed of PVA/CS, modified with properties after amending with cellulose nanocrystals (CNC), fractionated from waste obtained from acid-hydrolyzed rice straw (waste products of agriculture processes). The PVA/CS film was amalgamated with various concentrations of CNC which were 1, 3, and 5 wt%. The PVA/CS/CNC bionanocomposite film demonstrated excellent response as antifungal and antibacterial activity, a property that is mandatory and associated with potential films for the food processing industry [34]. PVA has a special interest in biocomposites as it has the ability to reduce antioxidant and antibacterial activities against gram-positive and gram-negative bacteria tested [35]. The mixing of PVA into chitosan was remarkably accepted as a new way to obtain films with promising biodegradability while retaining reasonable antioxidant and antibacterial properties. It provides balanced biocomposites with good strength, biodegradability, and application in extending the shelf life of packed food.

2.3 PVA/Starch-Based Biocomposites and Bionanocomposites

Starch is a natural highly biopolymer that is applied for the synthesis of biodegradable bionanocomposites of natural and synthetic polymers. The tructure of starch is composed of D-glucose, which is linked to form two homopolymer backbone chains: amylose and amylopectin. A linear α-D-(1,4)-glucan polymerizes to synthesize a linear structure of amylose. On the other hand, a highly branched chain known as amylopectin is similar to amylose but differs only due to branch points at α-1,6-linkages [36]. For composites made of starch, biodegrades by two distinct models are based on percolation theory. The first model focus on the behavior of degrading microbial species which make the contact with composite and invades through the surface of the composite. Meanwhile, the second model is based on enzymes' participation in the biodegradation of starch composites. Enzymes as macromolecular enter the composites through diffusion. Enzymes degradation generates small molecules that back-diffusion to the surface for direct uptake and assimilation by microorganisms [37]. Various research efforts have also been made toward the preparation of polyvinyl alcohol (PVA)/starch blended film, which can result in improved mechanical properties and biodegradability [38]. Mixing of starch with PVA in PVA/starch and barley husk-derived cellulosic composites increased the biodegradation in soil [39]. Nevertheless, the cross-linked polymer matrix ensures diminished biodegradation of film on account of structural complexity and increased hydrophobicity, which reduces further the contact angle and absorbance of water. Moreover, these composites have depicted 33% weight loss upon passing of 120 days of contact with soil. However, barley husk grafted biocomposite films had lower weight loss compared with the neat film. Higher amylose of maize starch can improve the biocomposite processability and biodegradation rates [40].

Maize Starch-(PVA)/clay nanocomposite films prepared through maize corn starch with the addition of 10 and 20 wt% of PVA (molecular weight, 13,000–23,000; hydrolysis degree, 87–89% and 3 and 5 wt% of Na-montmorillonite (Nanomer® PGV) are studied for biodegradation [41]. Gravimetrical tests and macroscopic observations helped to reveal the biodegradability of these nanocomposites. These films of 4 to 5 cm^2 and thickness of 0.1 to 0.2 cm placed at depth of 5 to 6 cm beneath the soil to assess the biodegradability. Polyethylene film biodegradation under similar conditions was used as a control under similar conditions for the comparison of maize starch-(PVA)/clay nanocomposite films biodegradation. Biodegradation is mainly controlled by maize starch. With the decrease in the maize starch contents, biodegradation rates were also reduced. However, the addition of glycerol, PVA, and MMT negatively impacts biodegradation in these nanocomposite films. Starch/PVA films losses were higher than pure PVA. Besides, maize starch, the addition of PVA, has been shown to reduce weight loss in nanocomposites. Native cornstarch and resistant starch/PVA composite have been synthesized using glycerol and citric acid as additives. Glutaraldehyde was used as a crosslinking agent in these composites. Glutaraldehyde improves the tensile strength but the elongation, swelling behavior, and solubility of glycerol and citric acid-added composites decreased [42]. Such properties improve the resistance to biodegradation in such composites. PVA/cassava starch biocomposites fabricated using bacterial cellulose fiber loadings. The role of cellulose fibers in improving moisture resistance and thermal properties has been noticed. For such changes in properties, strong hydrogen bonding of cellulose fibers with PVA/cassava biocomposites matrix has been observed [43].

Biodegradability has been assessed for PVA/*Trapa natans* starch (N-st) composite films. In such biodegradable films, PVA/N-st were grafted with methylmethacrylate (EMA) by applying a redox initiator—ferrous ammonium sulphate potassium persulphate. EMA has a positive effect on properties and a pronounced shift was noticed in properties such as the elongation at break, maximum stress, and reduction in water uptake in PVA/N-st-grafted-poly (EMA). These properties of composites showed that the biodegradation rate of natural PVA/starch-grafted-poly (EMA) composite film was slower than PVA/Starch [44]. To minimize the production costs of composites and decrease their unwanted impact on the environment, continuous efforts are underway in laboratories for preparing new cheap and green biocomposites. These hydrogels are reported to have potential applications in agricultural areas because of their good water absorption, excellent retention capacities, higher reswelling capability, and cost-effective and ecofriendly preparation. Semi-interpenetrating polymer network (semi-IPN) hydrogels of cassava starch-*g*-polyacrylic acid/natural rubber/polyvinyl alcohol composites have been developed with hydrogel properties. The semi-IPN hydrogel had confirmed 0.626 wt%/day biodegradation [45].

A study described the enzyme-based degradation of some montmorillonite mixed nanocomposites of PVA/starch and estimated mass loss and the reducing sugars. Montmorillonite as a nanoparticle hinders degradation of PVA/starch composites. Enzymatic degradation varies in the order as follows: PVA/starch/nanocore > PVA/starch/bentonite > PVA/starch/peruvian clay [46]. In events of erosion, the α-amylase adsorbs onto the biocomposites surface, and break down of α (1-4) glycosidic bond releases glucose which after that is followed by events of diffusion and further enzymatic degradation produces maltose also. Starch is an excellent source of the substrate to promote the activation of different natural microorganisms in the environment. Microorganism converts starch into low molecular weight metabolites, like ethanol, hydrogen, and methane under anaerobic conditions [47]. PVA is comparatively less biodegradable than starch [48]. This difference is more linked to microorganism preference for starch as the main source of energy for cellular processes. In the case of PVA, the main reason for less biodegradability is its synthetic nature. A specific microorganism is non-abundance and have enzymes that use PVA as substrate. The works of many laboratories have experimented with the decomposition of starch–PVA biocomposites/nanocomposites. Under compost conditions, rates of biodegrading of starch-PVA composites respond negatively to the amount of PVA, which degrades slowly [49]. Starch mediates the rapid biodegradation of films, which consists of both starch and PVA in comparison to films made up of pure PVA. Studies on starch glycerol samples revealed a loss up to 70% of initial dry weight basis after incubation in compost for 22 days [50]. This loss was minimized to 59% in starch–glycerol preparations after the amalgamation of PVA, signifying the addition of PVA decelerated the degradation process. The drawback in biodegradability of thermoplastic starch and PVA blends found decomposed residue enriched with the leftover-PVA while almost the entire degradation of starch occurred [51]. The PVA is linked to the stabilization rate of starch biodegradation in PVA/starch biocomposites.

Analysis of data of the biodegradability of the modified starch–PVA blends by applying bio-reactivity kinetic models predicted degradability of PVA was boosted after amending with starch [52]. The starch contents in PVA-starch blends determine the fast growth rate of degrading microorganisms under the first-order mode of reaction. The behavior of biodegradability of PVA, starch, and lignocellulosic films before and after cross-linking agent [53] was meticulously studied. Cross-linking portrayed a slow-down of the biodegradation of these films. The time required for film biodegradation was extended to 30 days in compost, with a loss of 50% to 80% mineralization. Pristine PVA and its blends which were cross-linked displayed relatively relaxed degradation and enhanced the influence of lignocellulosic fillers on structural losses of PVA in biocomposites under the biodegradation process. Biodegradability of nano-SiO_2 reinforced starch/PVA nanocomposite films with 5% nano-SiO_2 lost weight approximately 60%, almost equivalent to a weight loss of starch–PVA nanocomposites [54]. In some cases, nanoparticles may have a negative effect due to possible disruption of structure and poor biocompatibility of nanocomposites, which reduce the resilience against biodegradation.

2.4 PVA/Hemicellulose-Based Biocomposites and Bionanocomposites

Hemicellulose is the most abundant polymer, which is derived from renewable plant biomass. Hemicellulosic products have attracted the attention for the application of nontoxic, biocompatible, and biodegradable products. Nonstarch polysaccharides consist of xylans, mannans, and β-glucans with mixed linkages and xyloglucans and are commonly known as hemicellulose. These polysaccharides are chemically linked with cellulose in the cell wall of plants and extractable in aqueous alkaline solutions [55].

The degree of biodegradation of PVA/xylan composite film has continuously advanced within 30 days. Biodegrading species penetrating from the surface of the film into the inner core of the network could increase the probability of the breakdown of hydrogen bonds. The degree of biodegradation in these films was related to the composition of xylan. An increment of xylan composition could increase the disposal to microbial invasion correlated with the maximum degree of degradation from 33.5% to 56.0% after 30 days. Composite films without citric acid as a crosslinker had a higher level of biodegradation in contrast to composite films crosslinked with citric acid. Reactivity between xylan and PVA with 10% to 50% citric acid is initiated by stronger hydrogen and ester bonding among the chemical moieties during film formation [56].

In another work, composite of xylan/PVA (weight ratio of 3:1), xylans is hydrolyzed into xylooligosaccharides and xylose by xylanase in 7-hour enzymatic hydrolysis. These composites in the absence of urea had a higher degree of biodegradation as opposed to composite films carrying urea. Urea changes the biocomposite structure by enhancing the reaction of xylan and PVA. Urea decreased the enzymatic degradation degree. Enzymatic-mediated degradation of the pure PVA/xylan biocomposites peaked at maximum values as compared to similar the composite films mixed with glycerol or urea. This phenomenon explains that the additives into polymer networks altered the structural attributes of xylan in composition of composites. Moreover, unplasticized or urea-plasticized biocomposites had 99.25 and 95.28 % degree of degradation which was higher than that of glycerol or/and urea plasticized biocomposites degree of degradation of 93.46 and 92.80 % at 48 hours. Based on reported literature, PVA/xylan blending with butane tetracarboxylic acid (BTCA) has been widely used for biodegradability in soil. PVA/xylan + 10% BTCA had higher weight loss up to 41.29% as compared to PVA/xylan + 0% BTCA and PVA/Xylan 30% BTCA had a lower weight loss of 35.3% and 37.8%, respectively. Therefore, the addition of BTCA had a positive influence on the degradation of PVA/xylan blending films [57].

In another report, the biodegradability of composite films, PVA/xylan with glycerol and ammonium zirconium carbonate (AZC) was revealed in soil burial for 60 days. AZC influences the biodegradability of PVA/xylan composite by modification in water-resisting properties, water vapor permeability, solubility, mechanical properties, and thermal stability. Pristine PVA film exhibits nearly 10% degradation while the composite films revealed with higher value of 30% degradation under similar conditions. This degradation outcome provides the clue that PVA is attacked by microorganisms that are more specific and relatively not well known for their biodegradation activity in natural conditions [58]. Furthermore, the relatively high degradation amount of xylan in composite films makes it more susceptible to microbial attack. AZC reduces the impact of biodegradation in composite films as these films without AZC predicted as high as 32% degree of degradation. However, films with 10% AZC show with as high as 27% degree of degradation, which was less with a difference of 5% compared with composite films without AZC. The difference in the relative content of xylan in composite films and the inorganic substance (AZC) is ascribed to affect the composition due to AZC's hydrogen bonding with PVA/xylan protecting it against microbial attack.

Montmorillonite (MMT) is a common inorganic filler added into hemicelluloses for the purpose to develop hemicellulose/MMT films. PVA and chitin nano-whiskers (NCH), are incorporated into the hemicelluloses/MMT matrices to formulate hybrid films [59]. One feature of the nanocomposite hybrid films is their multidimensional characteristics, such as higher strength and oxygen barrier (due to electrostatic and hydrogen bonding). In addition, PVA and NCH are utilized in order to stimulate changes in the surface topology, thermal stability, transparency, oxygen barrier and mechanical strength for maintaining both the user-end desired properties and biodegradable functionality. Functional composite films originating from hemicellulose reinforced with PVA and nano-ZnO were developed. In these composite films, the solubility and processability of hemicellulose improved the functional properties.

Phosphatized PVAs esterified with phosphoric acid, which is used to fabricate hydrogels from hemicelluloses, PVA- phosphate (P-PVA), and chitin nanowhiskers. PVA and NCH added hemicelluloses/MMT hybrid films [60], Hemicellulose reinforced PVA, and nano-ZnO functional composite films [61] and cross-linked hemicelluloses, P-PVA, and NCH composite hydrogel [62] was assumed to be biodegradable but biodegradability has not been mentioned by the authors.

2.5 PVA/Polylactic Acid-Based Biocomposites and Bionanocomposites

An amorphous copolymer of PLA/poly(vinyl acetate-co-vinyl alcohol) (P(VAc-co-VA)) is reported as a solution to the problem of phase separation in PVA/PLA blends and diminished enzymatic hydrolysis of PLA in PVAc/PLA. Particular vinyl content of P(VAc-co-VA)/PLA determines tuning of miscibility and mechanical behaviour as compared with PVAc/PLA composite blends, for defining rates of enzymatic and nonenzymatic degradation as compared with PVA [63]. The difference in surface tension of pure PLA and 95/5 PLA/PVA blends showed a vast difference in their weight losses after degradation with proteinase K [64]. Blends of PLA/PVA (at 70°C, 15 hours) showed maximum degradation in comparison with aged-extruded PLA/PVA (at 2°C, 100 hours). Uniform degradation pattern found in de-aged samples and non-uniform degradation pursued in the case of aged samples to molecular weights and hydrolysis degrees of the of PVA in PLA/PVA blends made of ratios: 80/20, 90/10, and 97/3 wt.% can determine thermal and mechanical properties and crystallinity changes. An increase of PVA contents in PLA/PVA, it lowers the tensile strength for blends, especially at 20 wt.% of PVA, with 98% of hydrolysis degrees. Hybrid and blended nanofibrous poly (L-lactide-*co*-D, L-lactide) (PLDLLA) and PVA comprising of triclosan (Tri) has shown antibacterial activity with growth inhibitory zone diameter of 35 mm and 48 mm for studied bacterial strains, *E. coli* and *S. aureus*, respectively [65]. PVA accelerates the degradation rates of PVA/PLA blend films by intensifying the values of hydrophilicity and dropping the crystallinity of PLA. However, PLA/PVA blend film can be controlled to optimize hydrophilicity and degradability by carefully altering the ratio of PLA and PVA [66]. Crystallization kinetics and characterization of PLA and PVA blends with mass ratios of 100/0, 90/10, 80/20, 70/30, 60/40, 50/50, and 40/60 wt% were obtained, respectively [67].

The esterification reaction of glycerol with L-lactic acid covert it into a plasticizer called lacti-glyceride. which applied for thermal treatment of biodegradable PVA/PLA blend. Lacti-glyceride plasticization of PVA/PLA improves thermal and mechanical characteristics. Lacti-glyceride has a dual role in PVA/PLA blend as it acts as a plasticizer, as well as a compatibilizer. PVA/PLA blends plasticized with lacti-glyceride have much more hydrophobic surfaces than those blends plasticized with glycerol [68]. PVA/L-lactic acid (LA) biocomposites films showed bacteriostatic properties against bacterial strains i.e., *P. putida, S. epidermis,* and *Micrococcus* sp, under aquatic and aerobic conditions. In bacteriostatic activity against *P. putida, S. epidermis,* and *Micrococcus* sp. improved after the addition of LA into PLA. The addition of 40% of LA could reduce bacterial activity with an inhibition zone of a maximum of 350 mm. The pathway of biodegradation was triggered by esterified PVA and molecules of lactic acid in polymeric films [69]. The biodegradation of PVA/LA, with mixed microflora for 250 hours have been observed with a single lag phase appearing around 80 hours for pure PVA. LA was responsible for the two-phase biodegradation profile of PVA-LA. In the early phase of degradation, a preferential utilization of the easily degradable component, i.e., LA, is signified, whereas the later stages involve the degradation of the polymeric part in composites. LA mainly plays its role both as a plasticizer and esterification agent of PVA. The first phase of biodegradation showed maximum removal of total organic carbon due to non-bonded LA. LA grafted PVA backbone chain has caused a gradual decrease in kinetic constants with a step up in the content of LA in biocomposites. The restriction of lag phase < 30 h for PVA/LA50 than pure PVA has depicted the proof of chemical bonding between PVA and LA due to the presence of new centers which ease the potential microbial attacks.

2.6 PVA/Polyhydroxyalkanoates-Based Biocomposites and Bionanocomposites

Polyhydroxyalkanoates (PHAs) is a polyester, which is biosynthesized and accumulated inside the bacterial cell where it is stored as carbon material for energy production. However, when carbon sources are present in excess or growth conditions are unfavorable, these carbon materials are reduced for later use for energy generation [70]. The strength of PVA tied to PHA by grafting with maleic anhydride with the help of a dicumyl peroxide initiator doubled the strength of the PVA/PHA multilayer [71]. The biodegradation assays (ASTM 6691) have been used to study the biodegradation of PVA/PHA multilayer. The slow rate of mineralization was observed 84%

for PHA-*g*-MA compared with 97% for ungrafted PHA. PVA showed 12% of mineralization which determines the biodegradation resistance of PVA compared to PHA in the multilayer film. MA grafting of PHA hinders microbial attachment to some extent and results in the reduction of biodegradation from 97% to 84%. Biodegradable PVA/PHB blends revealed their properties are mainly affected by either PVA or PHB (if >30%). Hydrogen bonds developed by the hydroxyl group present in the PVA structure and ester group and hydroxyl functional groups of PHB are responsible for swelling and rupturing properties of the composite matrix. PHB could reduce the solubility of the blends as it is more hydrophobic than PVA [72].

PVA and PHB and their associated blends can be transformed into nanofibers through the application of the electrospinning technique. The crystallinity of PHB is suppressed in blends of nanofibers upon increasing the proportion of PVA. This change in crystallinity further accelerates the PHB degradation under the presence of increasing ratios of PVA [73]. [(*R*)-3-hydroxybutyrate-*co*-(*R*)-3-hydroxyhexanoate] (PHBH)/PVA blend nanofibers with ratios of 100/0, 90/10, 70/30, and 50/50 have shown biodegradation. Biodegradation has a difference in outcome due to the difference in compositions which arises due to intermolecular interaction in the amorphous phase, confirming the shift of hydroxyl and carbonyl stretching according to FTIR analysis. Contact angles increase with the increase of PVA ratio for nanofibers with PVA > 30% e.g., PHBH/PVA of 90/10, 70/30, and 50/50. These biodegradable nanofibers, *in vitro* enzymatic biodegradation using lipase solution at 37°C for 4 weeks demonstrated higher rates of biodegradation with an increase in PVA (>30%) as compared with pure PHBH and blend with 90/10 with no change over 1 week time (Figure 2.3). After a period of 4 weeks, weight loss was 5.1% and 11.9% for nanofibers of pure PHBH and blend 90/10, respectively, whereas weight loss of 18.3% and 20.5% reported for nanofibers made of PVA ratio of 30% and 50% PVA, respectively, after 4 weeks of lipase treatment. The degradation rate is influenced by nanofibers as it reduces the crystallinity of PHBH in presence of different loads of PVA. The mechanism is based on the diffusion of lipase solution inside nanofibers and removed the first amorphous region in the first phase of faster degradation and further proceeds slowly due to the crystalline region resisting the degradation of polymers chains [74].

In addition, the dissolution of PVA is faster than PHBH which possibly also promotes the enzymatic activity of lipase for further fragmentation of nanofibers. SEM micrographs shown in Figure 2.3, confirmed nanofibers

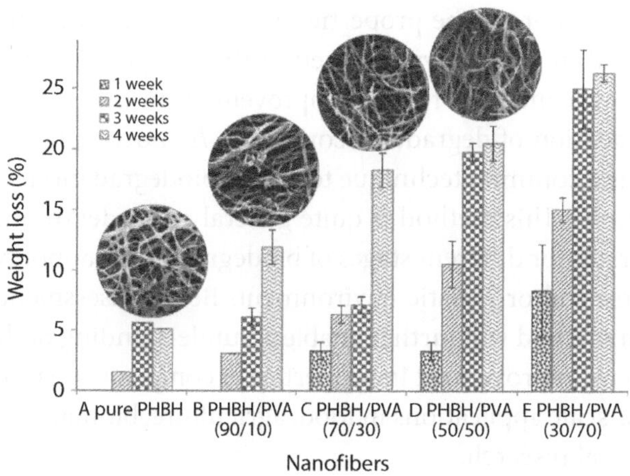

Figure 2.3 *In vitro* lipase driven degradation along with respective SEM micrographs for nanofiber blends composed of pure PHBH, different ratio of PHBH/PVA [75]. Reprinted after permission was granted from Elsevier.

of pure PHBH and PHBH/PVA 90/10 showed some broken nanofibers after cleavage occurs. PVA is associated with the damaging and curving of degraded nanofibers. In PHBH/PVA 50/50, nanofibers coalescence of fibers indicates the initiation of dissolution.

2.7 Conclusion

Polyvinyl alcohol possesses significant applications in the industry with the concern of its environmental pollution which require finding novel methods for the biodegradation of PVA. Subsequently, biocomposites biodegradation tests are limited to soil burial tests, which do not provide an understanding of the mechanism of weight loss. Therefore, a more detailed description on these methods needs to apply. Furthermore, biocomposites consisted of multiple components and possibly most difficulty exact reaching the mode of biodegradation. One component influence the biodegradation of other component used to synthesize the biocomposites. In addition, specific biodegrades are reported to enhance our understanding of the biodegradation of PVA biocomposites. Some PVA biocomposites are not well reported, and their biodegradation is less common like PVA/polylactic acid PVA/polyhydroxyalkanoate, PVA/hemicellulose. However, PVA biocomposites are considered important in packing material, and biocomposites improve the properties and make the biocomposites resistant compared to pure polymers used in the composition of biocomposites. This work concludes further improvement in methods for assessing the biodegradation of degradable composites/nanocomposites. Currently, weight loss is a common technique to report biodegradation, which has its own limitations. This method is quite general and a depth understanding of reaction rates for different stages of biodegradation cannot be calculated under diverse soil or aquatic environment. Better assessment of the biodegradation method will further enable an understanding of the structural and functional improvement in properties of composites/nanobiocomposites for large-scale applications in food, agriculture, biomanufacturing, and pharmaceutical research.

References

1. Kawai, F. and Hu, X., Biochemistry of microbial polyvinyl alcohol degradation. *Appl. Microbial. Biotechnol.*, 84, 2, 227, 2009.
2. Lucas, N., Bienaime, C., Belloy, C., Queneudec, M., Silvestre, F., Nava-Saucedo, J.-E., Polymer biodegradation: Mechanisms and estimation techniques–A review. *Chemosphere*, 73, 4, 429–442, 2008.
3. Kumar, G., Anjana, A.K., Hinduja, M., Sujitha, K., Dharani, G., Review on plastic wastes in marine environment – Biodegradation and biotechnological solutions. *Mar. Pollut. Bull.*, 150, 110733, 2020.
4. Li, C-x, Fan, X-r, Wang, Q., Jin, H., Effects of poly(vinyl alcohol) in desizing wastewater and its degradation performance. *Text. Aux.*, 6, 001, 2007.
5. Bing, X. and Yatong, X., Practice of treating PVA-bearing desizing wastewater. *Environ. Eng.*, 5, 7–9, 2002.
6. Du, G., Liu, L., Song, Z., Hua, Z., Zhu, Y., Chen, J., Production of polyvinyl alcohol-degrading enzyme with Janthinobacterium sp. and its application in cotton fabric desizing. *Biotechnol. J.: Healthcare Nutr. Technol.*, 2, 6, 752–758, 2007.
7. Raghul, S.S., Bhat, S.G., Chandrasekaran, M., Francis, V., Thachil, E.T., Biodegradation of polyvinyl alcohol-low linear density polyethylene-blended plastic film by consortium of marine benthic vibrios. *Int. J. Environ. Sci. Technol.*, 11, 7, 1827–1834, 2014.
8. Watanabe, Y., Morita, M., Hamada, M., Tsujisaka, Y., Studies on the polyvinyl alcohol-degrading enzyme. Part 1. Purification and properties of a polyvinyl alcohol-degrading enzyme produced by a strain of Pseudomonas. *Arch. Biochem. Biophys.*, 174, 573–579, 1976.

10. Yang, Y., Ko, T.-P., Liu, L., Li, J., Huang, C.-H., Chan, H.-C., Ren, F., Jia, D., Wang, A.H.-J., Guo, R.-T., Structural insights into enzymatic degradation of oxidized polyvinyl alcohol. *ChemBioChem*, 15, 13, 1882–1886, 2014.
11. Bian, H., Cao, M., Wen, H., Tan, Z., Jia, S., Cui, J., Biodegradation of polyvinyl alcohol using cross-linked enzyme aggregates of degrading enzymes from Bacillus niacini. *Int. J. Biol. Macromol.*, 124, 10–16, 2019.
12. Wei, Y., Fu, J., Wu, J., Jia, X., Zhou, Y., Li, C., Dong, M., Wang, S., Zhang, J., Chen, F., Bioinformatics analysis and characterization of highly efficient polyvinyl alcohol (PVA)-degrading enzymes from the novel PVA degrader Stenotrophomonas rhizophila QL-P4. *Appl. Environ. Microbial.*, 84, 1, e01898–01817, 2018.
13. Klomklang, W., Tani, A., Kimbara, K., Mamoto, R., Ueda, T., Shimao, M., Kawai, F., Biochemical and molecular characterization of a periplasmic hydrolase for oxidized polyvinyl alcohol from Sphingomonas sp. strain 113P3. *Microbiology*, 151, 4, 1255–1262, 2005.
14. Sakai, K., Hamada, N., Watanabe, Y., Degradation mechanism of poly(vinyl alcohol) by successive reactions of secondary alcohol oxidase and β-diketone hydrolase from Pseudomonas sp. *Agric. Biol. Chem.*, 50, 4, 989–996, 1986.
15. Jia, D., Yang, Y., Peng, Z., Zhang, D., Li, J., Liu, L., Du, G., Chen, J., High efficiency preparation and characterization of intact poly(vinyl alcohol) dehydrogenase from Sphingopyxis sp. 113P3 in *Escherichia coli* by inclusion bodies renaturation. *Appl. Biochem. Biotechnol.*, 172, 5, 2540–2551, 2014.
16. Alias, N.F., Ismail, H., Ab Wahab, M.K., Ragunathan, S., Ardhyananta, H., Ting, S.S., Development of new material based on polyvinyl alcohol/palm kernel shell powder biocomposites. *Adv. Environ. Stud.*, 2, 2, 98–107, 2018.
17. Olsén, P., Herrera, N., Berglund, L.A., Toward biocomposites recycling: Localized interphase degradation in PCL-cellulose biocomposites and its mitigation. *Biomacromolecules*, 21, 1795–1801, 2020.
18. Rathinavel, S. and Saravanakumar, S.S., Development and analysis of polyvinyl alcohol/orange peel powder biocomposite films. *J. Nat. Fibers*, 19, 1–10, 2020.
19. Kurniawan, A., Muneekaew, S., Hung, C.-W., Chou, S.-H., Wang, M.-J., Modulated transdermal delivery of nonsteroidal anti-inflammatory drug by macroporous poly(vinyl alcohol)-graphene oxide nanocomposite films. *Int. J. Pharm.*, 566, 708–716, 2019.
20. Mallakpour, S. and Khadem, E., Facile and cost-effective preparation of PVA/modified calcium carbonate nanocomposites via ultrasonic irradiation: Application in adsorption of heavy metal and oxygen permeation property. *Ultrason. Sonochem.*, 39, 430–438, 2017.

21. Liu, X., Chen, X., Ren, J., Chang, M., He, B., Zhang, C., Effects of nano-ZnO and nano-SiO_2 particles on properties of PVA/xylan composite films. *Int. J. Biol. Macromol.*, 132, 978–986, 2019.
23. Zhang, W., Yang, X., Li, C., Liang, M., Lu, C., Deng, Y., Mechanochemical activation of cellulose and its thermoplastic polyvinyl alcohol ecocomposites with enhanced physicochemical properties. *Carbohydr. Polym.*, 83, 1, 257–263, 2011.
24. Samzadeh-Kermani, A. and Esfandiary, N., Synthesis and characterization of new biodegradable chitosan/polyvinyl alcohol/cellulose nanocomposite. *Adv. Nanopart.*, 5, 01, 18, 2016.
25. Dalei, G., Das, S., Jena, S.R., Nayak, J., Samanta, L., Das, S.P., Surface modification of cellulose/polyvinyl alcohol biocomposites by non-thermal argon plasma: Applications towards biological relevance. *Cellulose*, 26, 4, 2437–2451, 2019.
26. Gooday, G.W., Physiology of microbial degradation of chitin and chitosan, in: *Biochemistry of Microbial Degradation*, pp. 279–312, Springer, Dordrecht, Netherlands, 1994.
27. Beier, S. and Bertilsson, S., Bacterial chitin degradation—Mechanisms and ecophysiological strategies. *Front. Microbial.*, 4, 149, 2013.
28. Kumar, A. and Kaur, H., Sprayed *in-situ* synthesis of polyvinyl alcohol/chitosan loaded silver nanocomposite hydrogel for improved antibacterial effects. *Int. J. Biol. Macromol.*, 145, 950–964, 2020.
29. Lisboa, H., Reinforcement of poly(vinyl alcohol) films with alpha-chitin nanowhiskers. *Polímeros (AHEAD)*, 28, 69–75, 2018.
30. Irvin, C.W., Satam, C.C., Carson Meredith, J., Shofner, M.L., Mechanical reinforcement and thermal properties of PVA tricomponent nanocomposites with chitin nanofibers and cellulose nanocrystals. *Compos. Part A: Appl. Sci. Manuf.*, 116, 147–157, 2019.
31. El Bourakadi, K., Merghoub, N., Fardioui, M., El Mehdi Mekhzoum, M., Kadmiri, I.M., Essassi, E.M., Qaiss, A.K., Bouhfid, R., Chitosan/polyvinyl alcohol/thiabendazoluim-montmorillonite bio-nanocomposite films: Mechanical, morphological and antimicrobial properties. *Compos. Part B: Eng.*, 172, 103–110, 2019.
32. Sofla, M.S.K., Mortazavi, S., Seyfi, J., Preparation and characterization of polyvinyl alcohol/chitosan blends plasticized and compatibilized by glycerol/polyethylene glycol. *Carbohydr. Polym.*, 232, 115784, 2020.
33. Khan, S., Garg, M., Chockalingam, S., Gopinath, P., Kundu, P.P., TiO_2 doped chitosan/poly(vinyl alcohol) nanocomposite film with enhanced mechanical properties for application in bone tissue regeneration. *Int. J. Biol. Macromol.*, 143, 285–296, 2020.

34. Perumal, A.B., Sellamuthu, P.S., Nambiar, R.B., Sadiku, E.R., Development of polyvinyl alcohol/chitosan bio-nanocomposite films reinforced with cellulose nanocrystals isolated from rice straw. *Appl. Surf. Sci.*, 449, 591–602, 2018.
35. Hajji, S., Chaker, A., Jridi, M., Maalej, H., Jellouli, K., Boufi, S., Nasri, M., Structural analysis, and antioxidant and antibacterial properties of chitosan-poly(vinyl alcohol) biodegradable films. *Environ. Sci. Pollut. Res.*, 23, 15, 15310–15320, 2016.
36. Nair, N.R., Sekhar, V.C., Nampoothiri, K.M., Pandey, A., 32 - Biodegradation of biopolymers, in: *Current Developments in Biotechnology and Bioengineering*, A. Pandey, S.N., C.R. Soccol (Eds.), pp. 739–755, Elsevier, Amsterdam, Netherlands, 2017.
37. Wool, R.P., Raghavan, D., Wagner, G.C., Billieux, S., Biodegradation dynamics of polymer–starch composites. *J. Appl. Polym. Sci.*, 77, 8, 1643–1657, 2000.
38. Das, K., Ray, D., Bandyopadhyay, N.R., Sahoo, S., Mohanty, A.K., Misra, M., Physico-mechanical properties of the jute micro/nanofibril reinforced starch/polyvinyl alcohol biocomposite films. *Compos. Part B: Eng.*, 42, 3, 376–381, 2011.
39. Mittal, A., Garg, S., Bajpai, S., Fabrication and characteristics of poly(vinyl alcohol)-starch-cellulosic material based biodegradable composite film for packaging application. *Mater. Today: Proc.*, 21, 1577–1582, 2020.
40. Chaudhary, A.L., Miler, M., Torley, P.J., Sopade, P.A., Halley, P.J., Amylose content and chemical modification effects on the extrusion of thermoplastic starch from maize. *Carbohydr. Polym.*, 74, 4, 907–913, 2008.
41. Katerinopoulou, K., Giannakas, A., Barkoula, N.-M., Ladavos, A., Preparation, characterization, and biodegradability assessment of maize starch-(PVOH)/clay nanocomposite films. *Starch-Stärke*, 71, 1–2, 1800076, 2019.
42. Yoon, S.-D., Chough, S.-H., Park, H.-R., Preparation of resistant starch/poly(vinyl alcohol) blend films with added plasticizer and crosslinking agents. *J. Appl. Polym. Sci.*, 106, 4, 2485–2493, 2007.
43. Abral, H., Hartono, A., Hafizulhaq, F., Handayani, D., Sugiarti, E., Pradipta, O., Characterization of PVA/cassava starch biocomposites fabricated with and without sonication using bacterial cellulose fiber loadings. *Carbohydr. Polym.*, 206, 593–601, 2019.
44. Kaur, K., Jindal, R., Maiti, M., Mahajan, S., Studies on the properties and biodegradability of PVA/Trapa natans starch (N-st) composite films and PVA/N-st-g-poly (EMA) composite films. *Int. J. Biol. Macromol.*, 123, 826–836, 2019.

45. Tanan, W., Panichpakdee, J., Saengsuwan, S., Novel biodegradable hydrogel based on natural polymers: Synthesis, characterization, swelling/reswelling and biodegradability. *Eur. Polym. J.*, 112, 678–687, 2019.
46. Spiridon, I., Popescu, M.C., Bodârlău, R., Vasile, C., Enzymatic degradation of some nanocomposites of poly(vinyl alcohol) with starch. *Polym. Degrad. Stab.*, 93, 10, 1884–1890, 2008.
47. Bai, F.W., Anderson, W.A., Moo-Young, M., Ethanol fermentation technologies from sugar and starch feedstocks. *Biotechnol. Adv.*, 26, 1, 89–105, 2008.
48. Gartiser, S., Wallrabenstein, M., Stiene, G., Assessment of several test methods for the determination of the anaerobic biodegradability of polymers. *J. Environ. Polym. Degrad.*, 6, 3, 159–173, 1998.
49. Chen, L., Imam, S.H., Stein, T.M., *Starch-polyvinyl alcohol cast film-performance and biodegradation*, American Chemical Society, Washington, DC (United States), 1996.
50. Mao, L., Imam, S., Gordon, S., Cinelli, P., Chiellini, E., Extruded cornstarch-glycerol-polyvinyl alcohol blends: Mechanical properties, morphology, and biodegradability. *J. Polym. Environ.*, 8, 4, 205–211, 2000.
51. Russo, M.A.L., O'Sullivan, C., Rounsefell, B., Halley, P.J., Truss, R., Clarke, W.P., The anaerobic degradability of thermoplastic starch: Polyvinyl alcohol blends: Potential biodegradable food packaging materials. *Bioresour. Technol.*, 100, 5, 1705–1710, 2009.
52. Chai, W.-L., Chow, J.-D., Chen, C.-C., Chuang, F.-S., Lu, W.-C., Evaluation of the biodegradability of polyvinyl alcohol/starch blends: A methodological comparison of environmentally friendly materials. *J. Polym. Environ.*, 17, 2, 71, 2009.
53. Cinelli, P., Chiellini, E., Gordon, S.H., Imam, S.H., Characteristics and degradation of hybrid composite films prepared from PVA, starch and lignocellulosics. *Paper Read at Macromolecular Symposia*, 2003.
54. Tang, S., Zou, P., Xiong, H., Tang, H., Effect of nano-SiO_2 on the performance of starch/polyvinyl alcohol blend films. *Carbohydr. Polym.*, 72, 3, 521–526, 2008.
55. Oliveira, E.E., Silva, A.E., Júnior, T.N., Gomes, M.C.S., Aguiar, L.M., Marcelino, H.R., Araújo, I.B., Bayer, M.P., Ricardo, N.M.P.S., Oliveira, A.G., Xylan from corn cobs, a promising polymer for drug delivery: Production and characterization. *Bioresour. Technol.*, 101, 14, 5402–5406, 2010.
56. Wang, S., Ren, J., Kong, W., Gao, C., Liu, C., Peng, F., Sun, R., Influence of urea and glycerol on functional properties of biodegradable PVA/xylan composite films. *Cellulose*, 21, 1, 495–505, 2014.
57. Gao, C-d, Ren, J-l, Wang, S-y, Sun, R-c, Zhao, L-h, Preparation of polyvinyl alcohol/xylan blending films with 1, 2, 3, 4-butane tetracarboxylic acid as a new plasticizer. *J. Nanomater.*, 2014, 1–8, 2014.

58. Chiellini, E., Corti, A., D'Antone, S., Solaro, R., Biodegradation of poly(vinyl alcohol) based materials. *Prog. Polym. Sci.*, 28, 6, 963–1014, 2003.
59. Chen, G.-G., Qi, X.-M., Li, M.-P., Guan, Y., Bian, J., Peng, F., Yao, C.-L., Sun, R.-C., Hemicelluloses/montmorillonite hybrid films with improved mechanical and barrier properties. *Sci. Rep.*, 5, 16405, 2015.
60. Chen, X-f, Ren, J-l, Meng, L., Influence of ammonium zirconium carbonate on properties of poly(vinyl alcohol)/xylan composite films. *J. Nanomater.*, 16, 1, 179, 2015.
61. Zhang, X., Luo, W., Xiao, N., Chen, M., Liu, C., Construction of functional composite films originating from hemicellulose reinforced with poly(vinyl alcohol) and nano-ZnO. *Cellulose*, 27, 3, 1341–1355, 2020.
62. Guan, Y., Qi, X.-M., Zhang, B., Chen, G.-G., Peng, F., Sun, R.-C., Physically crosslinked composite hydrogels of hemicelluloses with poly(vinyl alcohol phosphate) and chitin nanowhiskers. *BioResources*, 10, 1, 1378–1393, 2015.
63. Park, J.W. and Im, S.S., Miscibility and morphology in blends of poly(L-lactic acid) and poly(vinyl acetate-co-vinyl alcohol). *Polymer*, 44, 15, 4341–4354, 2003.
64. Gajria, A.M., Dave, V., Gross, R.A., McCarthy, S.P., Miscibility and biodegradability of blends of poly(lactic acid) and poly(vinyl acetate). *Polymer*, 37, 3, 437–444, 1996.
65. Ghaffari-Bohlouli, P., Hamidzadeh, F., Zahedi, P., Shahrousvand, M., Fallah-Darrehchi, M., Antibacterial nanofibers based on poly(l-lactide-co-d, l-lactide) and poly(vinyl alcohol) used in wound dressings potentially: A comparison between hybrid and blend properties. *J. Biomater. Sci. Polym. Ed.*, 31, 2, 219–243, 2020.
66. Wang, H., Mingang, S., Linfeng, Z., Yanhong L., Study on hydrophilicity and degradability of polyvinyl alcohol/polylactic acid blend film. *J. Biomed. Eng.*, 25, 1, 139–142, 2008.
67. Yeh, J.T., Yang, M.C., Wu, C.J., Wu, X., Wu, C.S., Study on the crystallization kinetic and characterization of poly(lactic acid) and poly(vinyl alcohol) blends. *Polym.-Plast. Technol. Eng.*, 47, 12, 1289–1296, 2008.
68. Li, H.-Z., Chen, S.-C., Wang, Y.-Z., Thermoplastic PVA/PLA blends with improved processability and hydrophobicity. *Ind. Eng. Chem. Res.*, 53, 44, 17355–17361, 2014.
69. Sedlařík, V., Saha, N., Kuřitka, I., Sáha, P., Preparation and characterization of poly(vinyl alcohol)/lactic acid compounded polymeric films. *Int. J. Polym. Anal. Charact.*, 11, 4, 253–270, 2006.
70. Choi, S.Y., Cho, I.J., Lee, Y., Kim, Y.-J., Kim, K.-J., Lee, S.Y., Microbial polyhydroxyalkanoates and nonnatural polyesters. *Adv. Mater.*, 32, 2020. https://doi.org/10.1002/adma.201907138

71. Thellen, C., Cheney, S., Ratto, J.A., Melt processing and characterization of polyvinyl alcohol and polyhydroxyalkanoate multilayer films. *J. Appl. Polym. Sci.*, 127, 3, 2314–2324, 2013.
72. Ol'khov, A.A., Iordanskii, A.L., Shatalova, O.V., Blends based on polyvinyl alcohol and polyhydroxybutyrate. *Int. Polym. Sci. Technol.*, 30, 8, 47–50, 2003.
73. Asran, A.S., Razghandi, K., Aggarwal, N., Michler, G.H., Groth, T., Nanofibers from blends of polyvinyl alcohol and polyhydroxy butyrate as potential scaffold material for tissue engineering of skin. *Biomacromolecules*, 11, 12, 3413–3421, 2010.
74. Cheng, M.-L., Chen, P.-Y., Lan, C.-H., Sun, Y.-M., Structure, mechanical properties and degradation behaviors of the electrospun fibrous blends of PHBHHx/PDLLA. *Polymer*, 52, 6, 1391–1401, 2011.
75. Rebia, R.A., Rozet, S., Tamada, Y., Tanaka, T., Biodegradable PHBH/PVA blend nanofibers: Fabrication, characterization, *in vitro* degradation, and *in vitro* biocompatibility. *Polym. Degrad. Stab.*, 154, 124–136, 2018.
76. Stoica-Guzun, A., Jecu, L., Gheorghe, A., Raut, I., Stroescu, M., Ghiurea, M., Danila, M., Jipa, I. and Fruth, V., Biodegradation of poly(vinyl alcohol) and bacterial cellulose composites by Aspergillus niger. *J. Polym. Environ.*, 19, 1, 69–79, 2011.
77. Lipsa, R., Tudorachi, N., Grigoraş, V. C., Vasile, C., Degradation of poly(vinyl alcohol)-graft-lactic acid copolymers by Trichotecium roseum fungus. *J. Appl. Polym. Sci.*, 132, 14, 2015.

3

Polyvinyl Alcohol-Based Bionanocomposites: Significance and Applications, Practical Step Towards Commercialization

S. Mohanapriya

EEC Division, CSIR-Central, Electrochemical Research Institute, Karailkudi, Tamilnadu, India

Abstract

Polyvinyl alcohol (PVA) is a nontoxic biodegradable, thermoplastic polymer with excellent thermo-mechanical properties with good reinforcing material property have gained interest in many applications in different fields especially biomedical, tissue engineering etc., PVA based Novel materials are essential for next generation bio-related applications. This chapter begins with a briefing about PVA emphasizing advantages of biocomposites and bionanocomposites describe appraise of a type of PVA based biocomposites and bionanocomposites and their suitability for biomedical and biomaterial applications. It then enumerate their suitability for prospects biomedical and bio-engineering applications. These ingenious materials are capable of function as biodegradable films, for membrane separation, for tissue engineering, for food packaging, for wound healing and dressing, hydro gels formation, gels formation, etc.. Nonetheless, core technologies of composite fabrication, high water absorption, system design, poor mechanical properties are perilous concerns connected with PVA based biocomposites and bionanocomposites.

***Keywords*:** Bionanocomposites, polyvinyl alcohol, biodegradable, composite fabrication, applications

Email: priyaechem@gmail.com

Visakh P. M. and Olga B. Nazarenko (eds.) Polyvinyl Alcohol-Based Biocomposites and Bionanocomposites, (59–80) © 2023 Scrivener Publishing LLC

3.1 Introduction: Polyvinyl Alcohol (PVA)

Polyvinyl alcohol, also known as PVOH, PVA, or PVAL, is a synthetic polymer that is soluble in water. It is effective in film forming, emulsifying, and has an adhesive quality. Polyvinyl alcohol is a resin, a natural or synthetic organic compound made of non-crystalline or viscous substance [1–3].

While other vinyl polymers are prepared by polymerization of its corresponding monomer, PVA undergoes either partial or complete hydrolysis of polyvinyl acetate to remove acetate groups.

Key raw-material to prepare PVA is the vinyl acetate monomer. The monomer is manufactured through the polymerization of vinyl acetate. Instead it goes through partial hydrolysis, which consists of partial substitution of the ester group with the hydroxyl group in vinyl acetate, completed in the presence of aqueous sodium hydroxide. The PVA is precipitated, washed and dried after gradual application of the aqueous saponification agent. When making the Polyvinyl alcohol solution, it is recommended to use tap water, as bacteria grows faster in PVA containing distilled water.

By annealing amorphous PVA films above 85°C, the glass transition temperature, semi-crystalline films of PVA were prepared. This allowed macromolecules to form crystallites, stabilizing the films and inducing a chemically cross-linked behavior. It has outstanding optical properties, great dielectric power, and excellent capacity for storing charges [4]. Doping with nanofillers can readily customize its mechanical, optical, and electrical attributes. PVA monomeric structure is depicted in Figure 3.1. Hermann and Haehnel first synthesized it in 1924 by saponifying the poly(vinyl ester) with a solution of sodium hydroxide resulting in a PVA solution [5]. PVA's physicochemical and mechanical properties are governed by the number of hydroxyl groups contained in the polymer PVA [6]. Different grades of PVA are available on the market based on hydrolysis (percent) and molecular mass and have different characteristics including melting point, viscosity, pH, refractive index and band gap [7]. The consequence of variation of the length and the degree of hydrolysis of vinyl acetate under

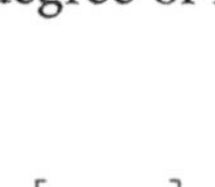

Figure 3.1 PVA structure.

acidic or alkaline conditions results in different PVA having different durability, tensile strength, density, emulsification extent, dispersing capacity etc., Presence of many hydroxyl groups on the PVA surface makes it one of the most hydrophilic polymers with high moisture sensitivity, and hence its resulting blends and composite materials have become popular for diverse range of applications. In specific, full-hydrolysis PVA is not known to be a thermoplastic polymer mainly because its melting temperature is very close to the temperature of degradation in the absence of plasticisers. Addition of plasticizers to PVA can decrease its melting temperature, brittleness and improve the flexibility and processability, on account of aggregate the segment mobility and reducing the crystallinity. Nevertheless, due to the increased hydrogen bonding between plasticizers and polymer molecules, the use of unnecessary plasticizers is likely to result in a phase separation. At the other side, partial-hydrolysis PVA produces groups of residual acetates, also known as vinyl acetate and vinyl alcohol copolymers. PVA was blended with either other polymers or inorganic materials, leading to formation of Nanocomposite or Composites that are more suitable for applications.

3.2 Properties of PVA

- ☐ Odorless,
- ☐ Non-toxic,
- ☐ Resistant to grease, oils, and solvents. It is ductile but strong, flexible, and functions as a high oxygen and aroma barrier,
- ☐ Good emulsifying and adhesive agents.

3.3 PVA Composites and Nanocomposites

Composites are heterogeneous multiphasic materials with improved properties as opposed to matrix phase. Figure 3.2 schematically represents steps involved in polymer composite preparation. The matrix phase is a continuous phase, while in fact the reinforcement stages are discontinuous. Throughout the mixed phases, individual components in the resulting composites are kept bound together by physical or chemical means while maintaining their physical identity.

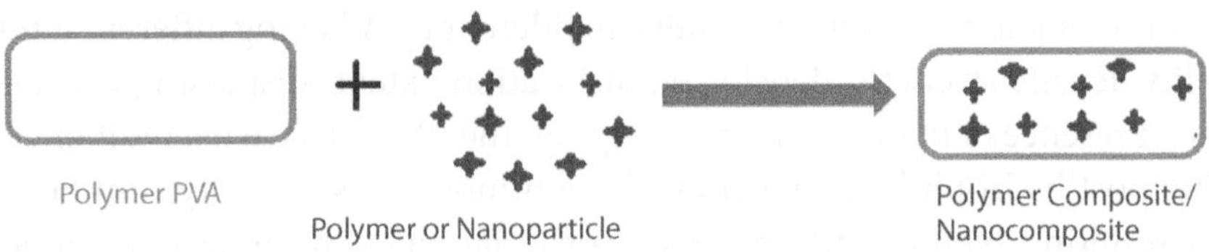

Figure 3.2 Schematics of PVA composite/nanocomposite preparation.

Attention to environmental friendly and full degradability has been moved due to sustainability and environmental impact problem. Conventional composite materials have micrometer-sized filler materials while nanocomposite materials have nanoscale filler materials, i.e. 1-100 nm. In the case of polymer matrix nanocomposites, polymer coils roughly have a diameter of about 50 nm, and there is strong contact between polymer and nanofiller. This will contribute to enhancement of nanocomposite properties. Developments of biodegradable material have risen over the past 10-12 years. It was found that neat PVA has fracture strength 34.1 MPa while strength goes on increasing with increase in the percentage of MFC up to 40% with the strength of 89.9 MPa.

In recent times much attention has been recorded in PVA as biocompatible, low cytotoxic and degradable polymer for biomedical and biomaterial research fields.

Nanocomposites are a fascinating class of materials in which nanomaterials in matrix material are reinforced and the nanomaterial has at least one dimension of up to 100 nm [8].

Many physical or chemical methods such as cross-linking, integrating new nanoparticles can be applied to further enhance their properties (such as mechanical properties, sensitivity to moisture). Using these new techniques, materials with a wide variety of property profiles could be developed and could even compete with conventional polymeric materials, both in price and performance.

Figure 3.3 shows different range of materials combined with PVA to fabricate PVA composite/nanocomposite. In recent years, PVA based polymer nanocomposites have been of great interest to the scientific and industrial communities, as these materials can greatly enhance mechanical properties. The production of new polymeric materials is primarily focused on the compounding, mixing and modification of a new polymer rather than on the chemical synthesis.

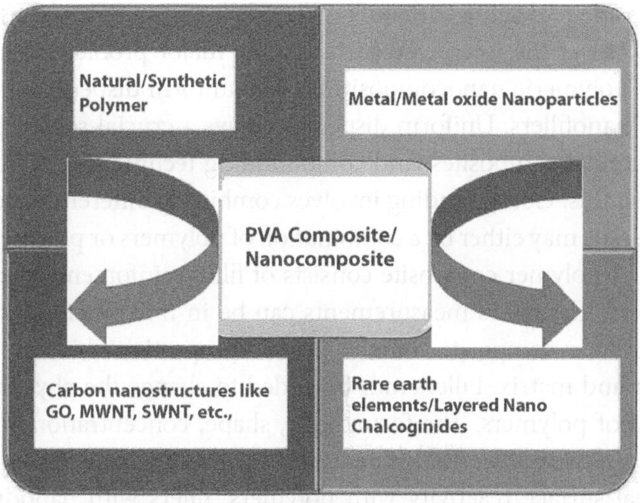

Figure 3.3 PVA composite/nanocomposite additive materials.

A wide variety of nanocomposites were prepared using PVA as matrix and Nano reinforcement, such as layered silicate. Researchers produced a series of thermoplastic starch/PVOH/montmorillonite micro- and nanocomposites which exhibit intercalated and exfoliated structures through extrusion processing. A small amount of PVOH (up to 7 wt%) and montmorillonite (up to 5 wt%) were used and the improvement of tensile strength (up to 67% increase) and tensile modulus (up to 85% increase) were reported [9]. They attributed the improvements in properties to both interfacial interactions of filler and matrix and the disruption of the recrystallization process. Starch/PVOH/montmorillonite composites by the melt mixing method is also studied. One of the indigenous application in order to minimize the concentration of PVA crystalline domains without harming its flexibility and mechanical stability over a wide temperature range. Such additives incorporate inert oxide ceramics, molecular sieves and zeolites, rare-earth oxide ceramics, solid super acids, ferroelectric materials and carbon. The effects of the inorganic additives have been analyzed in terms of Lewis acid-base interactions. When these effects are combined, the enhancing effect may be dominant or otherwise the impairing effect may be stronger. This might explain the different results of similar systems with nanoparticle with PVA.

PVA has a high melting point, due to hydrogen bonding in the matrix. Interfaces play a crucial role in understanding material behavior such as PVA.

One downside to dealing with bulk materials is the presence of a small fraction of atoms at the interfacial surface. One major problem in the manufacture of polymeric nanocomposites is the uniform dispersion in polymer matrix of nanofillers. Uniform dispersion plays a crucial role in producing multifunctional composites, and compounding techniques may be used to accomplish this. Compounding involves combining different materials, and such materials may either be a combination of polymers or polymer polymer additives. A polymer composite consists of filler reinforcement and matrix of polymers. The fillers measurements can be in micrometers or nanometres. A polymer composite contains certain properties which are absent in both filler and matrix. Fillers may be added to change the physicochemical properties of polymers. The form, scale, shape, concentration and mutual interaction have a powerful influence on the modification. Thanks to their exceptional surface reactivity with polymers, fillers with nanodimensions are used as smart dopant materials. Nanomaterials are usually graded into dimensions 0, 1, 2, and 3 (D). The materials of 0-D are in the form of quantum dots, nanolenses, nano onions. The nanowires, nanobelts, or nanotubes are present in the form of 1-D material.

Nanofillers (<5 wt percent) are most commonly used to enhance polymers' properties including durability, resilience, modulus, heat and scratch resistance, dimensional stability, thermal and electrical conductivity because of their high aspect ratio.

The main principle for improving the properties of polymeric nanomaterials is for interaction with surrounding matrix. Furthermore, nanofillers have a quantum containment effect that makes the interaction between nanofiller and polymer of great importance. PVA has been a suitable candidate for a number of applications over the last 50 years. Nanofillers made from metals, nonmetals, metal oxides, metal sulfides, inorganic or combinations of these materials can be used in PVA nanocomposites. Inorganic nanofillers may involve nanofibers of carbon, nanotubes of carbon, clays, silicates and others. PVA is used as an embedding mat and matrix for metal/inorganic sensing nanofillers, optoelectronic device design and many other applications.

3.3.1 Fabrication of PVA-Based Composites and Bionanocomposites

Nowadays, mobile electronic devices such as cell phone, laptop and personal digital assistant (PDA) etc., that are powered Li-ion batteries are largely being replaced with DMFCs, owing to high energy density and ease of operation/charging. Recharging battery needs time depending on type of electrochemical process on charging and recharging reactions. When charging process of battery takes place, electric current from outer power line results in electrochemical reaction at both anode and cathode in the battery. Due to occurrence of charge transfer process, ionic transport and mass transport combined with resistance components, it is complicated to amplify charging current of battery.

Operation of battery under higher overvoltage is risky since increase in overvoltage at higher load current densities restricts charging current. Most significant advantage of DMFC over batteries is prolonged usage time between recharges as well as elevated energy density than state-of-art batteries.

3.4 Categorization and Advantages of PVA Composites

Based on the modification method and modifying agent, it is classified as shown by Table 3.1.

Table 3.1 PVA commonly used modification and their applications.

Type of PVA systems	Modifying agent	Suitable application
Films/Membranes	Polymer/Metals/ Metal oxides/proton conducting agent/ Carbon nanostructure	Ion-conducting electrolyte
Hydrogel	Biopolymers/biomaterial	Tissue engineering, Drug delivery, etc.
Nanobiocomposite	Biopolymers/biomaterials/ Titanium oxide, Hydroxyl apatite etc.	Bio-implants, Orthopedic cartilage tissue, etc.
Biocomposite	Biopolymers/Chitosan/ Alginate/PEG etc.	Drug release kinetics, Bio-imgaing etc.

3.5 Issues Associated with PVA-Based Composites/ Nanocomposites

As the utilization of PVA is expected to rise universally the environmental threat and the industrial menace posed by PVA usage will discernibly increase. Presence of –OH (hydroxyl groups) in PVA result in high water up take, swelling and solubility in water which is one of the disadvantage of PVA film/composite. Due to water absorption the mechanical strength of the PVA film/composite decreases. On the other hand, some problems exist, such as worse processability and reduction of the ductility. The tailoring of the physicochemical properties of polymer nanocomposites is a major challenge when developing methodologies for synthesis. Systematic changes in experimental variables to explore causes and effects, strict control over experimental conditions using correct techniques, and strong basic knowledge about the parameters of the situation may help to resolve the challenges.

3.6 Diverse Applications of PVA-Based Composites/ Nanocomposites

3.6.1 Biomedical Applications

PVA nanocomposites may consist of nanofillers made from metals, nonmetals, metal oxides, metal sulfides, inorganic or arrangements of these materials. Because of the hydrophilic nature of PVA, crosslinkers are used to synthesize hydrogels for a number of applications. PVA is extensively used in biomedical applications for its compatibility, PVA composites, such as PVA gels, are used in diverse biomedical fields, such as in the engineering of contact lenses [10], artificial heart surgery [11], drug delivery systems [12] and wound dressings [13]. PVA may be used in a wide range of uses such as molecular sensing in biological and biomedical fields, membranes of fuel cells, chemo sensors, absorption of toxic metals, and optoelectronic devices. Nanofillers expose their larger surface area for interaction with polymer that is the important idea in the development of useful properties of polymeric nanomaterials. In medical devices, PVA is used as a biomaterial because of its extremely promising properties, for example biocompatibility, nontoxicity, non-carcinogenic, swelling properties, and bioadhesive features. This material is very valuable and desirable for biomedical application and uses. Excellent mechanical properties of the PVA nanocomposites were are intended to be composted at the end of their life rather than end up in landfills like most traditional petroleum-based non-biodegradable plastics. Normally used conventional modification reactions comprise acetalization, carbamation, etherification, or esterification of hydroxyl group. Due to its high chemical and thermal stability and low manufacturing costs PVA is used in many industries, such as garment, paper and food packaging industries. To make PVA fibers, numerous spinning techniques are being applied. Table 3.2 provides Implantable and Non-implantable devices biomedical uses of PVA. The PVA matrix plays a vital part in the formation of such valued membranes. PVA is being used to manufacture controlled interconnected micro porous three dimensional scaffolds. This is also a common water-soluble polymer, and has high water resistance and high optical clarity. PVA is used in labeling and as an optical

Table 3.2 Different applications of PVA and its modified form, product as implantable and non-implantable devices.

Device type	PVA form	Product	Patient contact
Non-Implantable Device	Polymeric Foam	Surgical sponge and Packing	Short Term Wound dressing
Non-Implantable Device	Polymer solution	Eye wetting Drops	Direct application to eyes
Non-Implantable Device	PVA Hydrogel	Contact Lenses	Direct application to eye tissues
Implantable Device	Hydrophilic coating	Polymeric Over coating	Long-term implant, blood contact
Implantable Device	Polymer hydrogel tubes	Nerve guides	Permanent implant, Nerve tissues
Implantable Device	Polymer hydrogel molded gel	Cartilage replacement	Permanent implant, blood and bone contact

polarizer because of these attractive properties. The biomedical usages of PVA contain embolization particles, eye drops, contact lenses, meniscus, artificial cartilage, and adhesion barrier.

Such material also occurs in other fields, including glass manufacturing, food packaging, sealing and binding, and adhesives. Graphite oxide/HA nanocomposites have many promising properties such as biocompatibility in medical science, are interested to develop biomaterials with coating capacity and having good mechanical strength. Oxide/HA based Poly(vinyl alcohol) (PVA) nanocomposite coating to study the feasibility of biocompatible coatings and hope this could Extensive use of PVA is observed in various scientific and technical fields, such as coating and finishing agents, emulsifiers, wood and leather.

3.6.1.1 Wound Dressing Material

Biomaterial scientists suggested, according to the first attempts, that hydrogels as dressing membranes meet all wound healing requirements and are suitable candidates for burning patients in the shortest possible time because of the following justifications: (1) hydrogels control the body's lost fluids and liquid and maintain wettability and moisture in the wound zone. PVA hydrogels extensively used for wound dressing. Hydrogels can absorb and retain the wound exudates which promote proliferation of fibroblasts and migration of keratinocytes. The last two processes are very important for full wound epithelialisation and cure. The tight mesh size of the hydrogels system additionally protects the wound from infection and prevents microorganisms and bacteria from entering the wound area. Previously the wet-retentive dressings were chosen based on the wound-shape form. Conversely, polymeric hydrogel dressings have been found to be comfortable at present for any form of wound and burn regardless of wound shape. The typical cotton-dressings of the gauze can re-injure the wound upon removal. Therefore, the occlusive polymer dressings were favored over the conventional and semi-occlusive desirable.

3.6.2 Cartilage and Orthopedic Applications

Because of its low protein adsorption properties, biocompatibility, high water solubility and chemical resistance PVA is widely used in medical devices. Some of PVA's most common medical uses are in soft contact lenses, eye drops, embolization particles, tissue adhesion barriers, and as cartilage and meniscus artificial. Due to their high water content, as well as their elastic and compressive mechanical properties, PVA hydrogels have been investigated for replacement of damaged cartilage. PVA cryogels used in cartilage resurfacing are made from small amounts of polymers of small molecular weight (generally 30 percent or higher PVA). Because of their rubber elastic physical properties, PVA hydrogels have been investigated as artificial cartilage replacements and because the hydrogels can be generated with tensile strength in the cartilage range of 1–17 MPa14 and compressive modules ranging from 0.0012 to 0.85 MPa depending on the polymer concentration and number of test cycles. Additional biocompatibility tests for PVA combined with other materials have been presented. Hydroxyapatite (HA), the bone's principal mineral base, was mixed with gelatin and PVA through emulsification to create a tissue engineering cartilage scaffold. This composite material in vivo by subcutaneously implanting it 12 weeks into the dorsal region of rats. Specific biomedical use for carboxymethyl chitosan and PVA combination has been considered as a drug delivery system implanted subcutaneously in rats, ensuing in high drug concentration retention and no cytotoxicity or hemolysis. The manufacturing process for PVA can be engineered to achieve the desired biomechanical properties. The thawing and freezing process, saline processing, crosslinking agents, and other materials all play a role in the final product's biomechanical properties. Fabrication and characterization of PVA/TiO_2 nanocomposite films [14] and PVA/Aminoacid composite [15] for orthopedic applications is demonstrated. Also, several investigators have documented PVA's wear features. The physical mixture of a polymer with the layered silicate may not form a nanocomposite. This situation is similar to polymer blends, and in most cases, separation into discrete phases may take place. In immiscible systems, which typically correspond to the more conventionally filled polymers, the poor physical interaction between the organic and inorganic components leads to poor mechanical and thermal properties.

3.6.3 Electrochemical Applications

Another promising and interesting area of developing PVA nanocomposite for electrochemical applications. Composite polymer electrolytes based on PVA can suggestively expand the performance in electrochemical devices such as lithium-ion batteries. Polymer blending is inexpensive and different method for attaining nanocomposites that optimizing the physical properties of innovative anticipated materials for diversity of applications. Conducting nanofibers of (PVA)-graphene oxide nanofiber coated with poly(3,4-ethylenedioxythiophene) for supercapacitor application was is also explored. Maximum ac conductivity of 2.49×10^{-3} S cm^{-1} is observed for 10 wt% TiO_2 doping level in PVA/CN-Li nanocomposite [16]. High-performance energy-storage devices, including supercapacitors, carry great capabilities for various applications, such as portable electronics, electric vehicles etc. Portable electronics industry includes the use of large-size and power-density energy storage devices. The supercapacitors need better morphology, surface area and carrier mobility of electrode materials. PVA-GO/PEDOT nanocomposite has also demonstrated higher activity with the energy density and power density of 9.58 Wh/kg and 304.37 W/kg, individually at 1.0 A/g current density. Multilayered ordered nanostructures fabricated by polyaniline nanowire arrays with graphene oxide nanosheets displayed energy density as high as 26.5 Wh/kg and the power density was ~3600 W/kg. PVA polymer in a straight line blended with nanosized MMT filler presented good ionic conductivity, thermal, and mechanical properties. The highest ionic conductivity value for the acidic PVA/10 wt.% MMT nanocomposite polymer membrane was around 0.0368 S cm^{-1}, which is suitable for fuel cell electrolyte. Fe3O4 nanoparticles functionalized CNTs incorporated into a PVA solution and used to prepare a cast film that exhibited open-circuit potential and a peak power density of 0.87 V and 87.8 mW cm^{-2}, respectively, were obtained using a 2 M methanol fuel in 6 M KOH at 60 °C with the PVA-based electrolyte containing 0.15% FeCNT.

PVA nanofibers introduced into Nafion in order to liaise in the proton conductivity of Nafion® displayed 0.022 S/cm at 70 °C and fully hydrated. Membranes containing PVA and poly(tetramethylene oxide) with heteropolyacid as a hydrophilic inorganic modifier were developed for low-temperature proton exchange membrane fuel cells showed a maximum conductivity of 4.8×10^{-3}S cm^{-1} at 80 °C. Provided their non-volatility, low toxicity, and high energy density, solid polymer electrolytes (SPEs) are an ideal alternative to liquid electrolytes. In this test, a SPE with an ion transport mechanism decoupled from a polymer's segmental movement based on poly(vinyl alcohol) containing salt lithium trifluoromethane sulfonate was prepared to resolve the low rates.

3.6.4 Optical and Photonic Applications

Because of their magnetic, electrical, and optical properties, metal NPs fall into an exciting research field which can be used in technical applications. Silver has the highest conductivity, both thermal and electric. Silver NPs possess potential uses in photonics, lighting, electronic part adhesives, thick film paste, and conductive inks. The water-soluble polymer polyvinyl alcohol has been chemically synthesized to make a composite material with cadmium selenide quantum dots at room temperature. In this paper, we examine photoluminescence spectra of CdSe/PVA nanocomposite solar cell, based on the absorber layer fabricated by spin coating synthesis route. The surface morphology of synthesized polymer nanocomposite film is studied by using scanning electron microscopy. Photoluminescence study shows the presence of quantum confinement effect in CdSe/PVA nanocomposite. The amorphous domains of semi-crystalline polymer allow the transport of ions, while improved polymer chains mobility leads to high ionic conductivity. Dielectric properties show the polymer chains' relaxation behavior as an important factor in the conduction of ions. In fact, the dielectric constant (ÿ) defines the polymer's ability to dissolve salt.

3.6.5 Renewable Energy Source-Based Applications

Energy is one of the most significant current topics of concern to human society. To substitute conventional fossil fuels, clean and renewable energies based on sunlight, wind, new chemicals, and biofuels are immediately necessitated. Accordingly, materials that can in a straight line transform or store renewable energies are being comprehensively studied. PVA based composites/bio composites are well-explored for various applications for energy generating devices and energy storing devices as well. They are capable of improving processability and/or flexibility of the materials and also generate fascinating interfacial properties with innovative functions. Up to the present time, various PVA based composites/bio composites with insulating or conducting polymers/carbon nanostructures have been prepared through noncovalent or covalent modifications. They act as vital components as electrode/electrolyte materials for lithium ion batteries, dye-sensitized solar cells, transparent conducting electrodes, catalytic electrodes, supercapacitors and low/high temperature operating fuel cells. Because of their excellent electrical properties and superior optical clarity the PVA/graphene materials have potential applications in microelectronics and solar cells as transparent conductive electrodes. Due to their broad applications in high-strength and conductive materials, catalysts, and energy-related systems, graphene/polymer composites have attracted considerable interest, especially flexible energy conversion and storage devices. Recently, a simplistic and scalable methodology to fabricate novel ABC-type terpolymer-based proton conducting membranes are prepared and elucidated the benefits of utilizing terpolymer composite membrane as an electrolyte for direct methanol fuel cells [17]. These PVA–TAF terpolymer composite membranes showed good thermal stability, flexibility, water uptake, and retention capacity. The prepared terpolymer composite membranes were thermally stable in a dry nitrogen atmosphere. Such membranes were studied and optimized based on proton conductance and their preliminary electrochemical properties (ionic exchange power, conductivity of protons, and swelling rate). In addition, spectroscopic techniques for positron annihilation were used to describe the sizes and concentrations of free-volume defects in the samples, which can significantly alter the physical and thermodynamical properties of PVA. Positron annihilation lifetime studies further indicated the effects of TAF incorporation into the structure and the configuration of structural defects in the polymer matrix. PVA/graphene composite has the potential to be used as a biocompatible and bio-safe conducting green composite. A PVA/graphene composite loaded with 0.5 vol.% of graphene had a conductivity of 5×10^{-1} S/cm and a maximum EMI shielding of ~19.5 dB (X-band) for a 1 mm thick film. S. Mo *et al.* prepared PVA/rGO nanocomposites with reduced graphene oxide loading of 50 wt.% and attained an electrical conductivity of $\sim 10^{11}$ orders higher than the pure polymer [18]. In solvent casting process, thin films of TiO_2/PVA nanocomposites were prepared to be used as UV-masks for various applications and as anti-reflective coatings in solar cells. The doping of TiO_2 nanoparticles into PVA matrix has had an effect on their semi-crystallinity and electronic structure. For this research TiO_2 nanoparticles have been used as nanofillers. This has a high refractive index, excellent opacity, hydrophilicity and good stability. Owing to the photocatalytic operation, it was used for water and air purification. In nature, TiO_2 is found in three phases: anatase, brookite, and rutile. The type of catalyst used in the counter electrode, light transmission, substratum conductivity, dye sensitizer, blocking substrate, and the band gap of the semiconductor used are among the many factors that affect the overall light to electricity conversion efficiency. DSSCs comprise a sealed assembly of dye-coated mesoporous semiconducting oxide photoanode nanoparticles covered with a redox molecules. Energy transfer in DSSCs starts when the dye's photo-excited electrons are inserted into the semiconducting substrate and eventually transferred to the counter electrode through the external circuit. At the counter electrode, the redox mediators return the electrons to the hole inside the dye molecule. The coating is regenerated and produces electricity. In solar cells, the use of liquid electrolytes makes the system vulnerable to leakage. Liquid electrolytes may contain volatile and corrosive components that can affect longevity of the solar panel. To solve these problems the researchers used gel or quasi solid polymer electrolytes. PVA based gel polymer electrolytes are widely studied for DSSC. Specific concentration of 1-butyl-3-methylimmidazolium iodide (BMII) ionic liquid has been added [19]. For the gel polymer electrolyte containing 20 wt. per cent, the highest room temperature ionic conductivity of 8.39 mS cm^{-1} was reported. The use of nanocomposite electrolytes instead of liquid and solid-state electrolytes is an easy and efficient way of improving photovoltaic efficiency as well as DSSC stability. There are specific physical, chemical, and electrochemical properties of each form of NF. They have their own effects on electrolyte output at DSSCs. Compared to using an individual polymer, polymer blending is a more attractive and simple method of enhancing properties such as biological, mechanical, and degradation. Novel PVA-Alginate based mixed-matrix membranes with heteropolyacids such as phosphomolybdic acid phosphotungstic acid and silicotungstic acid are reported for use as electrolytes in direct methanol fuel cells [20]. Hybrid nanocomposite PVA membranes prepared by integrating amino acid functionalized titanium dioxide bio-hybrid nanoparticles into a PVA matrix is also studied [21]. PVA/amino acid biocomposite membranes are explored as a new class of biocomposite membrane electrolytes for direct methanol fuel cells by Suganthi *et al.* [22]. Followed by this, Pectin blended with PVA in order to fabricate new class of hybrid nanocomposite followed by addition of sulfonated titanium dioxide (s-TiO_2) nanoparticles as inorganic proton conducting material [23]. PVA based electrolyte acted as a prototype of bio-functionalized membranes which rely on explanatory association between well-organized proton-based biological energy transformation process and proton conduction process in PEFC electrolyte under low humid operating [24]. For environmentally sustainable processing the combination of synthetic and natural polymers is favored. Synthetic polymers usually have a stable g relative to natural polymers.

3.6.6 Food Packaging Applications

PVA is the primary ingredient used in many end-use industries such as manufacturing, packaging, textiles, insulation, fibres, films, paper, printing, electronics, medicinal goods and food/beverages. Considering the value of food quality and health, a great deal of attention is given to creating modern active food packaging with antimicrobials and antioxidants. Antioxidants and antimicrobials minimize the oxygen content in the packaging and thereby slow the growth of microbes. Polyphenols derived from the tea leaves are among the most commonly used natural antioxidants. PVA is commonly used as a barrier film for foodstuffs and tablets for food supplements. It can shield foods from moisture, oxygen and other environmental components being exposed to them. Different grades of PVA such as high, moderate, and low viscosities are utilized in paper manufacturing. Exceptional properties like water resistance and high degree of swelling of PVA is very useful in the construction industry. Recently, PVA/Starch/Glycerol composite with good toughness and water absorption is shown to be potential film for packing food. Glycerol created a UV protective effect in the films, a significant outcome for food packaging to avoid lipid oxidative deterioration. Nanotechnologies that are currently being exploited for the development of intelligent packaging with enhanced communication function are presented, focusing mainly on oxygen, humidity and freshness indicators. Under the high humidity conditions applied in many packaging situations, the polymer matrix will need to be crosslinked properly to maintain its mechanical integrity. To this end, PVA has used freezing, heat treatment, irradiation and chemical crosslinking. Further important problems in food packaging are the low barrier properties to water vapor and methane. For example, fresh products that are alive (e.g., vegetables, fruits, meats) need to be packed with an optimum transmission rate in O_2 permeable materials, whereas refined goods do not require such mass transfer. Composite materials are those comprising two or more phases, one phase being the continuous phase and the other phase being the disperse phase. The continuous phase is usually a polymer, and the diffuse phase is the material for filling or strengthening. Before using nanocomposite materials for food packaging applications, the migration rate of nanocomposite elements to foodstuffs should be extremely measured due to the food safety issue. Many successful antimicrobial packaging solutions are based on the premise that they must release a known amount of antimicrobial molecules into the food matrix to exercise their positive action. These molecules should be trapped in a suitable matrix, suitable for keeping them within before use, but also for moving them when the matrix comes into contact with food. Sustainable packaging should meet the following criteria: (i) efficient, secure and sustainable for individuals and communities over their life cycle; (ii) market efficiency and cost requirements; (iii) use of renewable energy for processing, transportation and recycling; (iv) use of renewable or recycled source materials; (v) use of clean manufacturing technologies; (vi) use of materials. By virtue of their small dimension and surface reactivity, nanoparticles/PVA is capable to endow packaging materials with antimicrobial activity to inhibit the proliferation of spoilage and pathogenic microorganisms. Citric acid has proven to be an effective crosslinking agent for the development of a water-resistant PVA film, which is potentially ideal for active packaging. Despite the output caused by citric acid in the PVA virgin matrix, it is important to incorporate PEG into film formulation in order to diminish the material's hydrophilic behavior. Effect of incorporating different carboxylic acids namely oxalic acid (OA), succinic acid (SA) and citric acid (CA) on physicochemical and bioactive properties of PVA were investigated in detail for potential food packing [25]. PVA has been considered as one of attractive synthetic biopolymers over a few decades. PVA was blended with ST and GL to improve its biodegradability and compatibility whilst reduce material cost. PVA/ST/GL blend films had poor water resistance because all the components belonged to hydrophilic materials. Bio-degradable polymers like PVA have very low market value due to poor mechanical and physical properties. But, the biodegradability, low cost, and easiness of obtainability have insisted the researcher to improve their physical and mechanical properties by enhancing their functionality.

Also, many number of films namely, PVA/nanocellulose/Ag nanocomposite films [26], PVA/Starch [27], PVA/Chitosan [28], PVA/Gelatin [29], PVA/Alginate [30] etc., are actively investigated as potential antibacterial food packing materials. The most commercially available grades of PVA polymer are atactic, and the degree of crystallinity and physical properties depend on various factors, such as the manufacturing processes used to produce and transform the precursor polyvinyl acetate to PVA, the degree of hydrolysis, the molecular weight, the water content and the contents and types of plasticizers or other components incorporated. The active ingredients most widely used are salts or organic acids, such as potassium sorbate, ascorbic acid, benzoic acid, citric acid and alcohols. As the consumer demand for naturally extracted active agents is high, traditional sources are replaced by organic extracts. In order to improve the fresh-keeping performance, the antibacterial property of the films is very essential. PVA and carbon nanotube systems, the formation of nanotube induced crystalline polymer domains was the dominant reinforcement mechanism and no stress transfer to the nanotubes.

3.7 PVA Composites/Nanocomposites: Future Outlook

PVA is a synthetic polymer used in many medical and non-medical products over the last 30 years. PVA constitute the favorable class of synthetic polymer used in a variety of industries including biomedical, biomaterial, textile, adhesive, pharmaceutical and food industries and may cause environmental menaces. Multifunctional application of PVA based composite/Nanocomposite materials are attractive. Multiple non-clinical and clinical studies have shown that PVA is a synthetic alternative to the replacement of native cartilage, and is readily accessible compared to cartilage transplantation, which has restricted availability and transmission issues. Such problems include the refining of experimental methods, the strengthening of a systemic method and specialized knowledge contextual. As a result, substantial innovation in manufacturing and small- and medium-sized companies could be encouraged to grow. Hope the researchers can solve the problems and we can foresee an auspicious future for PVA composites. We conclude that PVA is a biologically active material that is stable *in vivo* (both in humans and in animals) and has appropriate biomechanical properties for potential tissue replacement implants to be a promising material.

References

1. Halima, N.B., Poly(vinyl alcohol): Review of its promising applications and insights into biodegradation. *RSC Adv.*, 6, 39823–39832, 2016.

2. DeMerlis, C.C. and Schoneker, D.R., Review of the oral toxicity of polyvinyl alcohol. *Food Chem. Toxicol.*, 41, 319–326, 2003.
3. Baker, M.I., Walsh, S.P., Schwartz, Z., Boyan, B.D., A review of polyvinyl alcohol and its uses in cartilage and orthopedic applications. *J. Biomed. Mater. Res. B Appl. Biomater.*, 100B, 1451–1457, 2012.
4. Kumar, A. and Ha, S.S., PVA-based hydrogels for tissue engineering: A review. *Int. J. Polym. Mater. Polym. Biomater.*, 66, 159–182, 2017.
5. W.O. Herrmann and W. Haehnel, U.S. Pat. 1, 672, 156, 1928.
6. Mok, C.F., Ching, Y.C., Muhamad, F., Osman, N.A.A., Hai, N.D., Hassan, C.R.C., Adsorption of dyes using poly(vinyl alcohol) (PVA) and PVA-based polymer composite adsorbents: A review. *J. Polym. Environ.*, 28, 775–793, 2020.
7. Nangia, R., Shukla, N.K., Sharma, A., Preparation, structural and dielectric properties of solution grown polyvinyl alcohol(PVA) film. *IOP Conf. Ser.: Mater. Sci. Eng.*, 225, 7–15, 2017.
8. Tang, X. and Alavi, S., Recent advances in starch, polyvinyl alcohol based polymer blends nanocomposites and their biodegradability. *Carbohydr. Polym.*, 85, 7–16, 2011.
9. Ali, S.S., Tang, X., Alavi, S., Faubion, J., Structure and physical properties of starch/polyvinyl alcohol/sodium montmorillonite nanocomposite films, agric. *Food Chem.*, 59, 12384–12395, 2011.
10. Hyon, S.H., Cha, W.I., Ikada, Y., Kita, M., Ogura, Y., Honda, Y., Poly(vinyl alcohol) hydrogels as soft contact lens material. *J. Biomater. Sci. Polym.*, 5, 397–406, 1994.
11. Jiang, H., Campbell, G., Boughner, D., Wan, W.K., Quantz, M., Design and manufacture of a polyvinyl alcohol (PVA) cryogel tri-leaflet heart valve prosthesis. *Med. Eng. Phys.*, 4, 269–277, 2004.
12. Kayal, S. and Ramanujan, R.V., Doxorubicin loaded PVA coated iron oxide nanoparticles for targeted drug delivery. *Mater. Sci. Eng.: C*, 30, 484–490, 2010.
13. Kokabia, M., Sirousazarb, M., Hassanc, Z.M., PVA–clay nanocomposite hydrogels for wound dressing. *Eur. Polym. J.*, 43, 773–781, 2007.
14. Mohanapriya, S. and Raj, V., Tuning biological properties of poly(vinyl alcohol) with amino acids and studying its influence on osteoblastic cell adhesion. *Mater. Sci. Eng. C*, 86, 70–82, 2018.
15. Mohanapriya, S., Mumjitha, M., Purna sai, K., Raj, V., Fabrication and characterization of poly(vinyl alcohol)-TiO_2 nanocomposite films for orthopaedic applications. *Mech. Behav. Biomed. Mater.*, 63, 141–156, 2016.
16. Rathod, S.G., Bhajantri, R.F., Ravindrachary, V., Sheela, T., Pujari, P.K., Naik, J., Poojary, B., Pressure sensitive dielectric properties of TiO_2 doped PVA/CN-Li nanocomposite. *J. Polym. Res.*, 22, 6–13, 2015.

17. Vasanthakumar, V., Mohanapriya, S., Priyadharsan, A., Anbarasan, P.M., Nambissan, P.M.G., Raj, V., Tunable physicochemical and free volume characteristics of novel terpolymer–poly(vinyl alcohol)-grafted membranes for direct methanol fuel cells. *New J. Chem.*, 43, 2942–2952, 2019.
18. Mo, S., Peng, L., Yuan, C., Zhao, C., Tang, W., Ma, C., Shen, J., Yang, W., Yu, Y., Min, Y., Epstein, A.J., Enhanced properties of poly(vinyl alcohol) composite films with functionalized graphene. *RSC Adv.*, 5, 97738–97745, 2015.
19. Teo, L.P., Tiong, T.S., Buraidah, M.H., Arof, A.K., Effect of lithium iodide on the performance of dye sensitized solar cells (DSSC) using poly(ethylene oxide) (PEO)/poly(vinyl alcohol) (PVA) based gel polymer electrolytes. *Opt. Mater.*, 85, 531–537, 2018.
20. Mohanapriya, S., Bhat, S.D., Sahu, A.K., Manokaran, A., Vijaykumar, R., Pitchumani, S., Sridhar, P., Shukla, A.K., Sodium-alginate-based proton-exchange membranes as electrolytes for DMFCs. *Energy Environ. Sci.*, 3, 1746–1756, 2010.
21. Mohanapriya, S., Rambabu, G., Suganthi, S., Bhat, S.D., Vasanthkumar, V., Anbarasu, V., Raj, V., Bio-functionalized hybrid nanocomposite membranes for direct methanol fuel cells. *RSC Adv.*, 6, 57709–57721, 2016.
22. Suganthi, S., Mohanapriya, S., Raj, V., Bio composite proton exchange membrane electrolytes for direct methanol fuel cells. *J. Appl. Polym. Sci.*, 133, 43514, 2016.
23. Mohanapriya, S., Rambabu, G., Bhat, S.D., Raj, V., Pectin based nanocomposite membranes as green electrolytes for direct methanol fuel cells. *Arabian J. Chem.*, 13, 2024–2040, 2020.
24. Mohanapriya, S., Rambabu, G., Bhat, S.D., Raj, V., Hybrid membranes for polymer electrolyte fuel cells operating under various relative humidity values. *J. Solid State Electrochem.*, 21, 3437–3448, 2017.
25. Suganthi, S., Mohanapriya, S., Kanaga, V.R.S., Vignesh, R.D.S., Sundar, J.K., Tunable physicochemical and bactericidal activity of multicarboxylic-acids-crosslinked polyvinyl alcohol membrane for food packaging applications. *ChemistrySelect*, 3, 11167–11176, 2018.
26. Wang, W., Yu, Z., Alsammarraie, F.K., Kong, F., Lin, M., Mustaph, A., Properties and antimicrobial activity of polyvinyl alcohol-modified bacterial nanocellulose packaging films incorporated with silver nanoparticles. *Food Hydrocolloids*, 100, 105411–25, 2020.
27. Liu, B., Xu, H., Zhao, H., Liu, W., Zhao, L., Li, Y., Preparation and characterization of intelligent starch/PVA films for simultaneous colorimetric indication and antimicrobial activity for food packaging applications. *Carbohydr. Polym.*, 157, 842–849, 2017.
28. Pereira Jr., V.A., Queirozde Arrud, I.N., Stefani, R., Active chitosan/PVA films with anthocyanins from Brassica oleraceae (red cabbage) as

time–temperature indicators for application in intelligent food packaging. *Food Hydrocolloids*, 43, 180–188, 2015.

29. Meshram, J.V., Koli, V.B., Phadatare, M.R., Pawar, S.H., Anti-microbial surfaces: An approach for deposition of ZnO nanoparticles on PVA-gelatin composite film by screen printing technique. *Mater. Sci. Eng.: C*, 73, 257–266, 2017.
30. Shahidul Islam, M. and Karim, R., Fabrication and characterization of poly(vinyl alcohol)/alginate blend nanofibers by electrospinning method. *Colloids Surf. A: Physicochem. Eng. Asp.*, 366, 135–140, 2010.

4

Polyvinyl Alcohol/Cellulose-Based Biocomposites and Bionanocomposites

Nor Asikin Awang[1,2], Mohamad Azuwa Mohamed[3]* and Wan Norharyati Wan Salleh[1,2]†

[1]*Advanced Membrane Technology Research Centre, Universiti Teknologi Malaysia, Skudai, Malaysia*

[2]*Faculty of Chemical and Energy Engineering, Universiti Teknologi Malaysia, UTM Skudai, Johor Darul Ta'zim, Malaysia*

[3]*Department of Chemical Sciences, Faculty of Science and Technology, Universiti Kebangsaan Malaysia, UKM Bangi, Selangor, Malaysia*

Abstract

Synergistic interaction between polyvinyl alcohol (PVA) and cellulose-based material has emerged as a new opportunity to fabricate high-performance bio-composites and bio-nanocomposite material for various potential applications. These polyvinyl alcohol and cellulose-based materials have unique chemical and physical properties,as well as facile preparation of its bio-composites and bio-nanocomposite material that received huge attention among scientists around the globe. Thus, this chapter mainly recapitulated and reviewed various kinds of PVA/cellulose-based bio-composites and bio-nanocomposites and their promising preparation techniques. On top of that, the critical physicochemical properties of the selected materials have been deliberated via various characterization techniques. Finally, the usage of PVA/cellulose-based bio-composites and bio-nanocomposites in the promising application and its performances are highlighted and discussed.

Keywords: Polyvinyl alcohol, cellulose, fabrication, characterization, emerging application

**Corresponding author*: mazuwa@ukm.edu.my

†*Corresponding author*: hayati@petroleum.utm.my

Visakh P. M. and Olga B. Nazarenko (eds.) Polyvinyl Alcohol-Based Biocomposites and Bionanocomposites, (81–130) © 2023 Scrivener Publishing LLC

4.1 Introduction

Over recent years, we have witnessed ample development in the quality of life, consequently from the excogitation and development of non-renewable sources called petroleum-based resources. However, day by day, these sources are being depleted and have become severe issues. Addressing the crisis of the petroleum-derived polymers increased the interest in the utilization of "green polymers," which can be originated from natural resources. The term "green polymers" refers to the natural, sustainable polymers, which are biodegradable, environmentally friendly and can be reused with minimum energy consumption. Yearly, around 35 million tons of natural fibers have been harvested as being reported according to the Food and Agricultural Organization (FAO) [1]. Natural polymers generally can be obtained from many resources, which can be categorized based on their properties, chemical structures and also origins of the sources. Generally, this natural polymer can be naturally found in grasses, leaves, stalks of plants, and animals. The natural polymers are majorly found with the presence of naturally cellulose contents. Other than cellulose, the components of the polymers are also made up of lignin, hemicellulose, pectin, pigments and other derivatives. Cellulose is the most abundant form of natural polymers with an inestimable quantity of raw material. Out of 700 000 billion tons of global quantity of cellulose, only 0.1 billion tons is presently consumed for the manufacturing process of paper, textiles, pharmaceutical compounds and any other products for consumers [2].

Chemically, the structure of cellulose consists of long-chain repeating β-D-glucose sugar units, which are linked together. The elimination of water molecules created the linkage of the sugar units. The glucose units present in the cellulose polymers are called anhydroglucose units. The cellulose having three different types of anhydroglucose units named as the reducing end with a free hemiacetal group located at C-1, nonreducing end with a free hydroxyl at C-4 and the internal rings joined at C-1 and C-4. The linkage of two molecules of sugar units creates disaccharides which are called cellobiose. Referred to the structure of the cellulose, the glucose units are located in six-membered rings called pyranoses [3]. These pyranoses structures are linked together via single oxygen atoms, which are named acetal linkages. The linkage connects the C-1 of one pyranoses ring to the C-4 of the neighbour ring. The formation of acetal formed due to the removal of the water molecules that occurred between an alcohol group and the hemiacetal group. The structure of the cellulose is illustrated in Figure 4.1.

The types of cellulose can be differentiating based on the configuration, chemical composition and properties of the cellulose. The cellulose material can be classed into various types, including cellulose fiber, cellulose nanofiber, cellulose acetate, bacterial cellulose, regenerated cellulose, cellulose nanocrystalline, and cellulose hydrogel. The natural fibers that had a higher percentage of cellulose possess excellent mechanical strength but easily to be more flammable. For augmentation reasons, the cellulose is extracted from natural fiber and applied for the production of composite, owing to the hierarchical structure and semicrystalline nature. Notwithstanding their advantages, the cellulose sourced from the natural fiber found some difficulties in terms of converting them to different types of configuration as it is not easily dissolved in the common solvent. This involvedness occurs due to the nature of the cellulose, which has strong intermolecular hydrogen bonding, possess a high degree of polymerization and also high crystallinity degree [4]. Conjointly with the environmentally safe guideline, the biodegradable polymer, which is polyvinyl alcohol (PVA) preferably reinforced to cellulose-based materials in order to create eco-sustainable composites [5].

PVA is an interesting polymer having good chemical and physical properties, which poses high crystallinity, biodegradable, water-soluble, high thermal stability with non-toxic properties [6–8]. In addition, PVA also has good film-forming capability due to the presence of the hydroxyl groups, which strengthen the intermolecular hydrogen bonding of the structure [6, 9, 10]. The PVA can be synthesis by simple preparation via hydrolysis of polyvinyl acetate. Owing to the good properties with relatively low cost of production, PVA is majorly applied in most of the applications related to drug delivery [11–15], tissue engineering scaffolds [16–21], packaging

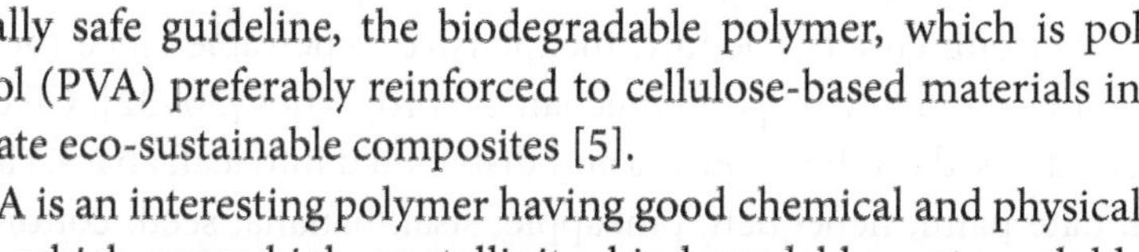

Figure 4.1 The chemical structure of cellulose is made up of repeating glucose units [1].

[22–24], gas separation [25–27], coating [28, 29], and etc. The biocompatibility of the cellulose-based composites combining with the PVA will produce highly preferable materials to be matched with the extensive applications in the industries. The applications of these merging composites can be broadened far beyond the conventional applications areas. Undoubtedly, PVA/cellulose-based biocomposites and bio-nanocomposites are exceptionally vital for humankind. Thus, in this chapter, we reviewed the topical developments, detailed ways of preparations, characterization techniques of composites and the recent potential applications in various fields of industries.

4.2 Polyvinyl Alcohol/Cellulose-Based Biocomposites and Bionanocomposites and Their Preparation

4.2.1 Polyvinyl Alcohol/Cellulose Fibers

Cellulose fiber (CF) is being used as potential reinforcing material because it possesses many advantages such as being plentifully available, low weight, environmentally friendly, inexpensive, renewable since these are waste biomass and having good mechanical properties [30–32]. CF can be categorized based on their sources and distributed into leaf: abaca, cantala, curaua, date palm, henequen, pineapple, sisal, banana: seed: cotton bast: flax, bagasse, jute, ramie: fruit: coir, kapok, oil palm: grass: alfa, bagasse, bamboo: stalk: straw. Although the CF owned many benefits, the properties of the CF are affected by some factors, including the climate, harvest time, maturity, pre-treatment process, types of modification and technical processing. According to Box and colleagues, flax fiber is mentioned as a strong fiber compared with other types of fiber, which has tensile strength up to 1500 MPa that has been observed at 3 mm clamp length [33]. Among the listed CF sources, the bast and leaf types are majorly utilized in composite applications. Plant fiber is composed of CF that build-up of cellulose microfibrils (CMF) connected together by an amorphous lignin matrix. The structure of the natural fiber cell is illustrated in Figure 4.2.

However, CF also has some drawbacks such as moisture adsorption, quality variation, low thermal stability and meagre compatibility with hydrophobic polymer matrices [3]. Thus, the blending of the CF with biodegradable polymers, which is PVA, is widely done to produce new PVA/

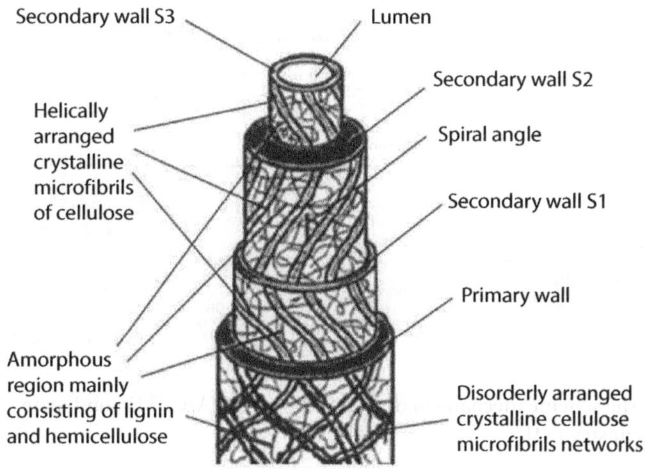

Figure 4.2 Structure of natural cellulose fiber [3].

CF with better properties that can be tally with the recent application in the industrial fields. A mechanochemical method via solid-state mechanochemistry is one of the ways to enhance the dispersion of the fabricated PVA/CF [34]. Mechanochemical technique refers to the stress-induced reaction based on the mechanics and chemistry process. Compared to the common solvent-based chemical reaction, this method is believed to be an energy-saving process and gentle environmental technique. Fabrication of PVA/CF, which derived from the hardwood pulp fibers, was successfully cross-linked with the presence of the succinic anhydride (SA) as a cross-linker [34]. The mechanism of the fabrication of PVA/CF is presented in Figure 4.3.

In another work, *Grewia optiva* fiber was reinforced in PVA/starch to form biodegradable starch/PVA/CF composite films via a cross-linking process by using citric acid as plasticizer and glutaraldehyde as the cross-linker [35]. *Grewia optiva* fiber with the dimension of 15 to 20 μm homogeneously mixed with starch/PVA by using a mechanical stirrer before being dried to form flat sheet films. Next, another approach of the utilization of PVA impregnated with CF nonwovens is prepared via two techniques [36]. CF nonwovens are the typical CF in which the fibers are held together mainly by mechanical entanglement. The first method is called the applicator method (AM), and the second method is the calendar method (CM). During the AM process, the polymer solution is placed on the nonwoven

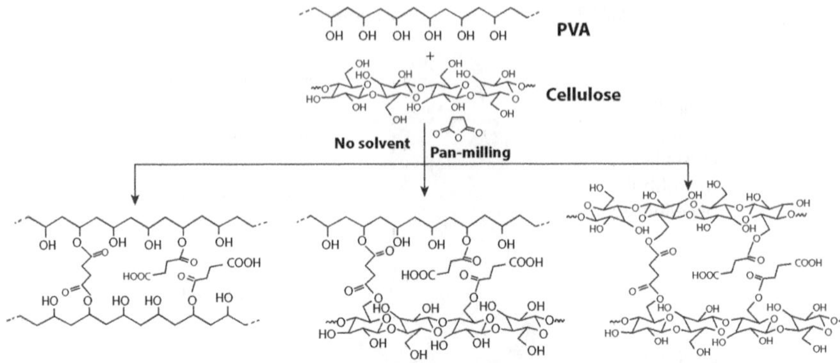

Figure 4.3 The scheme mechanism of the reaction of PVA/CF via solid-state mechanochemistry [34].

by casting method and afterward drawing of the applicator along with the sample. Meanwhile, in CM approaches, the impregnated PVA/CF is squeezed once between elastomer-coated calendar rolls.

4.2.2 Polyvinyl Alcohol/Cellulose Acetate

Cellulose acetate (CA) having lots of advantages, especially to be commercialized as the membrane materials in membrane separation technology. Many of the publications from the previous researchers highly highlighted the capability of the CA in many fields of the applications due to its high biocompatibility, high potential flux, affordable price of material and ease of availability [37–40]. The CA, with a very good mechanical strength and having a great moisture management property, make it suitable to be implemented in biomedical applications such as tissue engineering and antibacterial applications [41–44]. Besides, the CA, which is the derivative of cellulose, own less amount of hydrogen bonding compared with the pristine cellulose, thus carrying ample movement of the molecular chains to achieve higher crystallinity [45]. The suitability of the CA to act as the polymer matrix is related to its ability that it can be easily moulded into different forms, including membrane, fibers and spheres. The nature of CA, which having high hydrophilicity properties, enhanced the accessibility of the aqueous solution to the surface of the hybrid materials such as PVA [31]. However, especially in the fabrication of membrane, the water-like nature of the CA disturbed the diffusion of the non-solvent (water) and hindered the coagulation process during the phase inversion that will produce a dense skin layer with low permeate flux. Thus, the modification of the CA with PVA was found to be able to limit the poor characteristics of the CA [46].

The fabrication of PVA/CA membrane can be completed via the common phase inversion method [46, 47] and the casting method [48]. Being different, the preparation of CA membrane coated with PVA and Fe_3O_4 nanocomposites was done via *in situ* formations of Fe_3O_4 in the polymeric solution containing PVA through the facile chemical procedure [49]. The produced nanocomposites PVA/Fe_3O_4 were prepared by dissolving PVA in water and *in situ* formation of magnetite nanoparticles with a certain additional amount of Fe_3O_4. Therefore, the prepared CA membrane prepared via the phase inversion method was immersed in PVA/Fe_3O_4 before being dried for 2 h at 60 °C.

4.2.3 Polyvinyl Alcohol/Bacterial Cellulose

Bacterial cellulose labelled as BC has been considered as a biomaterial with a dazzling future in many industrial fields, especially in medical due to its biocompatibility, biodegradability, high clarity, having fine fiber network structure, high surface area, high crystallinity with high holding water capacity and possess excellent strength-to-weight ratio [50–56]. The development of the BC was swiftly observed in order to widen the application of BC in industrial's demand. As a progress, high interest was also paid to the lessening of BC production costs by substituting with cheaper materials, industrial waste or by-products. The production of cellulose by aerobic bacteria as an extracellular polysaccharide membrane is a bottom-up approach that finally produces highly pure cellulose. In contrast, the top-down process in the production of cellulose involving the extraction of cellulose sourced from plants and woods [54, 55]. The production of BC can be produced from bottom-up approaches as a protection against UV radiation or chemical environment and to allow the accessibility of oxygen from the atmosphere [57]. The sources of BC can comes from various types of bacteria including Acetobacter [51, 58], Agrobacterium [59, 60], Pseudomonas [61–63], Rhizobium [64–66], Sarcina [67], and Komagataeibacter xylinus [68–70]. The synthesis of the BC was firstly synthesized from bacteria called Aceto*bacter xylinum* that was initially reported in 1886 by Brown [71]. The 3-D network of cellulose fibrils obtained from *Acetobacter xylinum* illustrated in Figure 4.4.

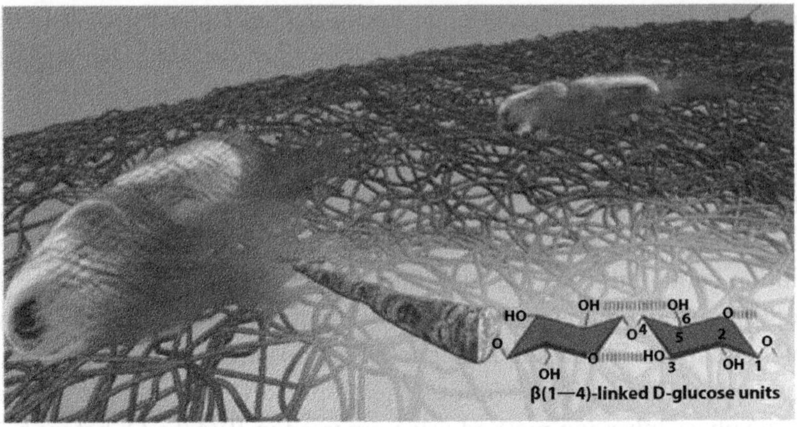

Figure 4.4 Well-arranged 3-D design of BC obtained from *Acetobacter xylinum* [72].

Though the chemical composition of BC is the chain molecules linked by cellobiose as same as other cellulose, it is free of contaminant molecules such as lignin, hemicellulose and pectin. Therefore, the purification of the BC can be directly obtained and can be highly purified with NaOH solution with minimal energy needed. However, BC is combined with PVA in order to improve the physicochemical properties and expand its potential to be matched with the industrial field's demand. The preparation of PVA/BC can be prepared via many techniques. Previously, the BC was compounded with PVA colloidal sol to form physical crosslinking hybrid hydrogel [73]. The physical hydrogel can be attained by repeating the cycle of freezing and thawing, which results in the formation of crystallites and gelation [17]. This type of technique will leave out the additional chemical cross-linker. In addition, physical hydrogel offers better biocompatibility and show a good mechanical property due to the crystalline areas and the stronger interlinkage between the molecules that able to give improved cross-linked network link, which directly increases the mechanical load or stress. Specifically, the prepared PVA/BC was frozen at −20°C for 12 hours. After a definite time, the PVA/BC was taken out from the freezer and thawed for 2 to 3 hours at room temperature condition before being frozen again up to five cycles of freezing and thawing process. The effect of the thawing and freezing process onto PVA/BC is illustrated via the SEM images in Figure 4.5. The addition of BC turned some discontinuous pores of PVA into regular and uniform distribution porous structures.

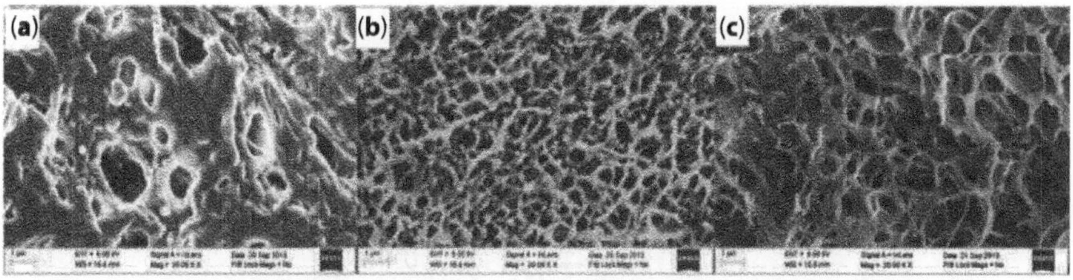

Figure 4.5 SEM images of (a) pure PVA; (b) PVA/BC with 1:30 and (c) PVA/BC with ratio 1:10 [73].

The increased ratio of PVA/BC from 0:1 to 1:30 resulting a loose and less uniform with a pore diameter of PVA/BC of about 0.2–2.0 μm. By the similar freezing-thaw method, which is specifically in artificial cornea biomaterial, the composite hydrogel of PVA/BC yielded new properties, including having high water content, high visible light transmittance and suitable UV absorbance, improved mechanical strength, and possess high thermal stability [74].

The effect of sonication steps in the fabrication of PVA/starch blended with BC was reported by Abral and groups [75]. Ultrasonication method was able to minimize the viscosity of the biopolymer and improved the filler dispersion in the matrix [76–78]. Compared to unsonicated PVA/BC, the sonicated PVA/starch blended with BC produced less viscous, with the starch gel became shorter and higher in mobility resulting more homogeneous structure of the blend. In another approach, the *in-situ* growth process was done by the direct addition of PVA into *Acetobacter xylinum* inoculated medium and compared with the composites done by impregnation of BC gels with PVA [79]. As illustrated in Figure 4.6, the *in-situ* method produced optically transparent nanocomposites films while having excellent mechanical properties and tougher compared to pure PVA, pristine BC film and impregnated PVA/BC, which is suitable in packaging application [80]. Besides, the technique in preparing PVA/BC can be prepared via the dry-casting method for antimicrobial tested against *Escherichia coli* [81]. BC powder was dispersed into the PVA solution under vigorous stirring for 1 h at room temperature before being poured into the Petri dish and allowed to dry. A similar technique was applied by Indriyati and co-workers in order to cast PVA/BC [82]. Another casting method in preparing PVA/BC was prepared through the solvent casting method [83].

Figure 4.6 Physical observation of (a) pure BC; (b) *in situ* BC; (c) Impregnated BC and (d) *in situ* PVA/BC [79].

4.2.4 Polyvinyl Alcohol/Regenerated Cellulose

Regenerated cellulose (RC) is a natural polymer manufactured by the conversion of natural cellulose to cellulose derivative by regeneration, physical dissolution and shaping process that extensively researched due to its several desirable properties such as renewable sources, biodegradability, high transparency, low cost of material and high versatility owing to its ability to be modified easily [84–87]. The properties of RC, which possess high transparency and inborn biocompatibility, make RC become the main material in electronic components, consumer products and aerospace in the near future [88]. Naturally, the presence of the inter- and intra-molecular bonding between the hydrogen bonding in the cellulose structure limits the solubility of the cellulose in the common solvents [89]. Thus, the dissolution systems that are able to increase the dissolution rate of cellulose is widely studied [90–92]. Previously, reported dissolution solvent systems have been researched, including the ionic liquid systems, N-methylmorholine N-oxide ($NMMO.H_2O$) [86, 93] LiCl/N, N-dimethylacetamide, NaOH-urea [89, 94, 95] and tetrabutylammonium fluoride/dimethyl sulfoxide system. The dissolution and regeneration techniques are a "green" and the clean process leading to sustainable development and environmental preservation. However, as mentioned by Mohamed and groups, the NaOH-urea dissolution system is minimal toxic production and a less expensive method, and it has been considered environmental friendly to dissolve the cellulose [89, 96, 97]. The dissolution of recycled newspaper, RNP in 7 wt% NaOH/12 wt% urea to form transparent regenerated cellulose membrane (RCM) was successfully fabricated to be potentially utilized in membrane separation and filtration system [89].

However, the properties of the RC can be mechanically and physically improved by incorporating with the polymer such as PVA. The modification of the RC can be done by the most simple method involving the immersion of the RC film in aqueous solutions containing the modifying agents. For example, the composite film of chitosan/PVA/RC dissolved from MCC is prepared by immersion of the RC film in a bath containing PVA for 1 hour before being dried for 48 hours. Environmental friendly food packaging produced from PVA/RC composite film with the addition of glycerol as the plasticizer is prepared by simple immersion into PVA bath for 1 hour at the different concentrations of PVA as proposed in Figure 4.7. The prepared PVA/RC is cut into a specific size and placed in the desiccators for 5 days before being used for the test [98].

RC extracted from cotton linter successfully blended with PVA in ionic liquid 1-allyl-3-methylimidazolium chloride (AMIMCI) in coagulating water bath [99]. The formation of the film is cast on the glass plate before being coagulated in the distilled water. The transparency of the fabricated PVA/RC compared with RC is demonstrated. As observed, the wrinkle of RC film is obvious compared to PVA/RC film, indicating that the shrinkage

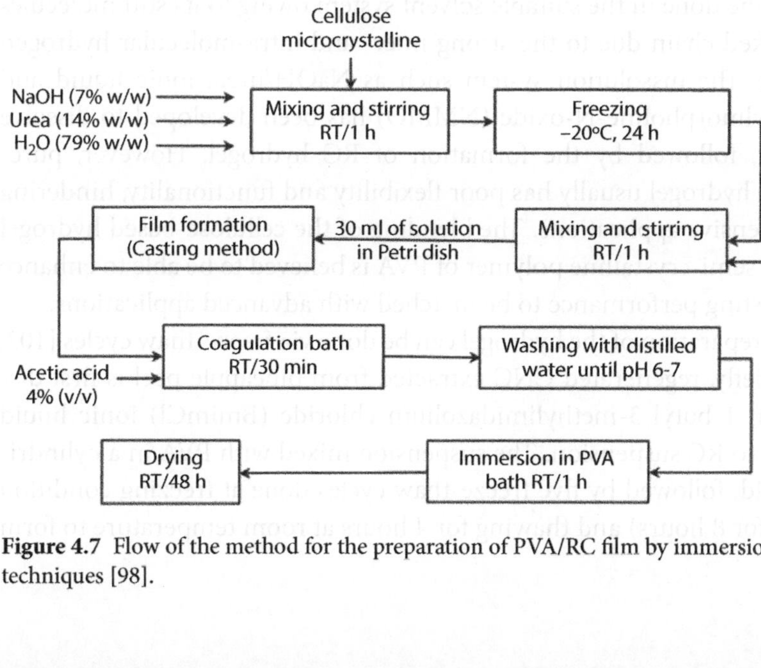

Figure 4.7 Flow of the method for the preparation of PVA/RC film by immersion techniques [98].

of RC is greatly improved compared to the RC incorporated with PVA, which is consistent in the mechanical properties. Similar work was done by Lu and co-workers in which the dissolution and the blending of PVA/RC derived from microcrystalline cellulose (MCC) prepared in AMIMCI [100, 101]. Interestingly, the transparent film made up from RC derived from Kenaf core powder that is dissolved in LiOH/urea solvent reinforced with PVA successfully fabricated and can be reprinted repeatedly several times [102].

4.2.5 Polyvinyl Alcohol/Cellulose Aerogel or Hydrogel

The hydrogel can be defined as a three-dimensional network of hydrophilic polymer created from the chemical and/or physical cross-linking and has the capability to absorb a large amount of water and biological fluid [103]. The remarkable properties of the hydrogel, which has the high swelling ability and designable great responsiveness, hydrogel are widely applied in agriculture, tissue engineering [17, 104], drug delivery [105] and wound dressing [106]. Although synthetic polymer-based hydrogel possesses many advantages, in some situations, there is a limitation of the application due to its lack of biocompatibility. In contrast, cellulose biopolymer-based hydrogel is more practical to be applied in industrial fields due to its attractive properties. The preparation of cellulose-based hydrogel needs to be done in the suitable solvent system owing to its stiff molecules and packed chain due to the strong inter- and intra-molecular hydrogen bonding. The dissolution system such as NaOH/urea, ionic liquid and N-methylmorpholine-N-oxide (NMMO) has been developed to dissolve cellulose, followed by the formation of RC hydrogel. However, pure-cellulose hydrogel usually has poor flexibility and functionality, hindering their extensive applications. The blending of the cellulose-based hydrogel with the semi-crystalline polymer of PVA is believed to be able to enhance their existing performance to be matched with advanced applications.

The preparation of the hydrogel can be done via freeze-thaw cycles [103, 107]. Briefly, regenerated CNC extracted from pineapple peel is first dissolved in 1-butyl-3-methylimidazolium chloride (BmimCl) ionic liquid solution to RC suspension. The suspension mixed with PVA in a cylindrical mould, followed by five freeze-thaw cycles done at freezing condition (−20°C for 8 hours) and thawing for 4 hours at room temperature to form

PVA/CNC hydrogel. Additionally, fabricated PVA/CNC hydrogel is magnetized with the Fe_3O_4. The preparation process of the PVA/CNChydrogel with the addition of magnetic nanoparticles is schematically illustrated in Figure 4.8.

A series of cellulose hydrogels from Whatman CF-11 fibrous cellulose powder physically and chemically cross-linked with PVA in NaOH/urea aqueous dissolution system is fabricated [108]. The preparation of hydrogel from the dissolved PVA/cellulose solution is done by repeating freezing/thawing cycles up to 7 times. After 7 cycles, the cellulose/PVA mixed solution dried at 60°C for 12 hours to form gels. Recently, by using the same freezing/thawing method, tricaroxy cellulose (OxC) was incorporated with PVA [109]. OxC mixture composites and pure PVA aqueous solution are subjected to 3 freezing-thawing cycles. After each of the freezing cycles, the OxC/PVA is kept at 4°C to ensure the formation of the porous structure. Next, the preparation of PVA/CNC/poly(2-Hydroxyethyl methacrylate) (PVA/CNC/poly HEMA) and PVA/CNC/poly(N-methylenebisacrylamide) (PVA/CNC/polyMBA) hydrogels prepared by photo-crosslinking followed by freezing/thawing cycle [110]. The process of the preparation of PVA/CNC/poly HEMA and PVA/CNC/polyMBA is detailed in Figure 4.9.

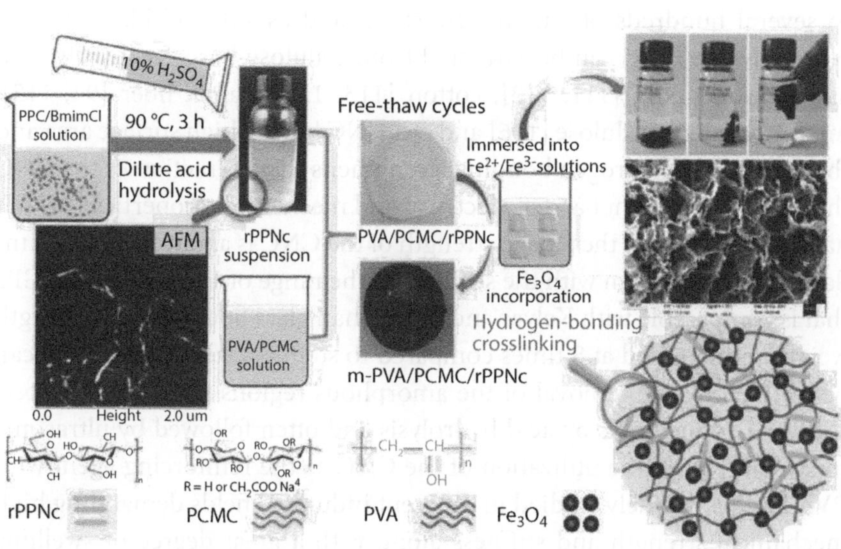

Figure 4.8 Preparation flow of PVA/CNC hydrogel with the additional magnetic properties of Fe_3O_4 [103].

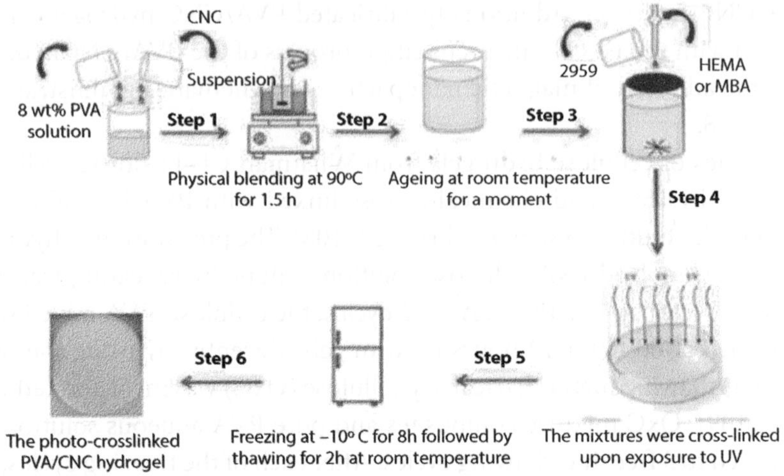

Figure 4.9 Preparation of PVA/CNC/poly HEMA and PVA/CNC/polyMBA hydrogel via photo-crosslinking followed by freezing/thawing cycle [110].

4.2.6 Polyvinyl Alcohol/Cellulose Nanocrystals

Cellulose nanocrystals (CNC), similarly mentioned as nanocrystalline cellulose (NCC) having a width of 2 to 30 nm which can be lengthened to several hundreds of nanometers structured as stiff rod-like particles. Normally, the CNC can be extracted from cellulose-based materials such as hardwood pulp [111, 112], cotton [113, 114], kapok fiber [94, 115], microcrystalline cellulose [116] and etc. CNC having high surface area and the presence of hydrogen bonding strengthens the interaction of the CNC that gave an excellent barrier, electrical and mechanical properties [117]. It has been found that the tensile strength of the CNC is about 500 MPa, similar to the aluminium with the stiffness in the range of 140 GPa to 220 GPa that is comparable with Kelvar and better than glass fiber with the strength weight ration noted at 8 times compared to stainless steel [118]. CNC can be purified by the removal of the amorphous regions from the cellulosic sources raw materials by acid hydrolysis and often followed by ultrasonic treatment [115]. The utilization of the CNC as the reinforcing agent with PVA has been widely studied in different industrial fields demanding high mechanical strength and stiffness along with a great degree of swelling. In addition, CNC can form a dense network owing to strong inter fibrilar bonds that beneficially be used as barrier materials [119]. The dimensional of the CNC structure results in a high surface area, consequently creating powerful interaction between CNC and the matrices to produce well-constructed nanocomposites.

The reinforcement of CNC in the polymeric materials is found able to increase the stiffness and the tensile strength of the composite by reducing the elongation at break. Reinforcement of the CNC sourced from bleached kraft softwood pulp with PVA is done by homogeneously blended with cellulosic paper to improve the strength properties of paper via improving inter-fibers bonding [120]. As cited by Babu *et al.* the isolation of CNC from rice straw purified through acid hydrolysis process composited with PVA/chitosan was successfully reinforced by the solution casting method [121]. The purification of the CNC from rice straw via the acid hydrolysis process is digitally photographed in Figure 4.10.

Other than rice straw, the CNC is extracted from kiwi *Actinidia deliciosa* pruning residue by applying leaching process and subsequently by acid hydrolysis process [24]. The resultant CNC is used as the reinforcement phase in PVA that is further blended with chitosan. Fabricated PVA/CNC produced from cotton linter prepared from two-stage of the process: (1) primary hydrolysis to remove the amorphous regions of cellulose and (2) fragmentation of the crystalline region to produce CNC was done by Noshirvani and co-workers [122]. The generated PVA/CNC shows some

Figure 4.10 Preparation of the CNC derived from the rice straw via acid hydrolysis process [121].

improvement in mechanical and barrier properties suggested that it can be a good replacement for mineral reinforcement such as silica.

4.2.7 Polyvinyl Alcohol/Cellulose Nanofiber

Cellulose nanofibers (CNF) refer to the fibrilar unit that is produced from the linear combination of cellulose macromolecules which having both amorphous and crystalline regions. The combination between them will create an entangled network. The extraction of CNF involved mechanical, chemomechanical, and enzymatic techniques [123, 124]. Among three of the common ways of extraction, enzymatic and chemomechanical proved as minimal energy consumption required compared to mechanical approaches. The extracted CNF had a high aspect ratio and figured web-like structure due to the presence of the hydrogen bonding among the fibers and also because of the mechanical interlocking of the CNF. The incorporation of the CNF in polymeric materials with a diameter of CNF less than 100 nm is believed to be able to be fabricated that will directly improve the mechanical properties of the created CNF-polymeric materials. This fact is related to the properties of CNF, the having very high stiffness up to 134 GPa for cellulose crystal and possess strength up to 10 GPa with a high specific surface area [125]. Previously, as mentioned by Kakroodi, the reinforced CNF with a composite polymer having moderate mechanical performance owing to the strong hydrophilic nature of CNF, which limit the interfacial adhesion with hydrophobic polymers [126]. Additionally, the –OH groups that created hydrogen bonds in CNF directly created agglomeration between the fibers, which reduced the specific surface area of CNF. Thus, very high energy is needed in order to defeat the agglomerated CNF and created a good dispersion of the CNF in the formed matrices solution. Consequently, the extraction of CNF via the chemomechanical method is usually kept in the form of an aqueous suspension. Subsequently, the composite is prepared via solution casting using a water-soluble polymer, which is PVA as the matrix.

The fabrication of CNF sourced from *aloe vera* (*Aloe barbadensis* Miller) reinforced with PVA via solvent casting technique was successfully fabricated [126]. The transparent and flexible PVA/CNF with a thickness approximately of 200 μm is obtained after the addition of PVA with different concentrations of CNF ranging from 2 wt% to 10 wt% is illustrated in Figure 4.11. The slightly yellow colour of the PVA/10% CNF is due to the

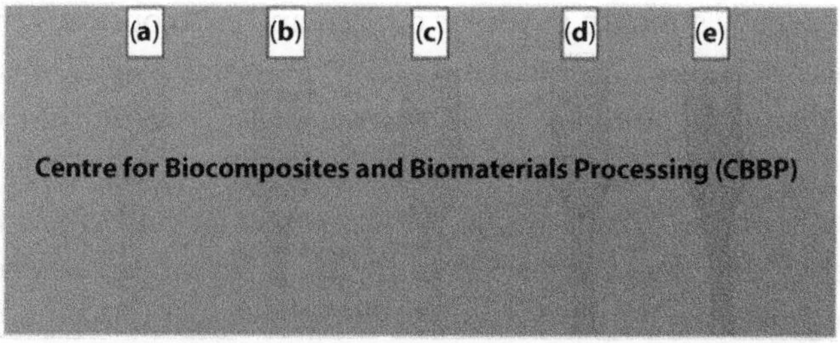

Figure 4.11 Optical observation of the transparency of the fabricated nanocomposites consists of (a) PVA/0% CNF; (b) PVA/2% CNF; (c) PVA/5% CNF; (d) PVA/7% CNF and (e) PVA/10% CNF [126].

presence of the residual lignin that is not fully separated during the isolation of the CNF from the *aloe vera.* The same technique in the fabrication of PVA/CNF has also been reported [9, 127]. Fabrication of CNF sourced from MCF derived from Alfa grass *Stipatenacissima* reinforced into PVA by exfoliation is demonstrated by Ben Cheikh and groups [128]. Next, the CNF, which is purified from flax fiber and commercialized microcrystalline cellulose (MCC), was reinforced with PVA to form a casted composite, proved both of the PVA/CNF (sourced from flax fiber and MCC) able to

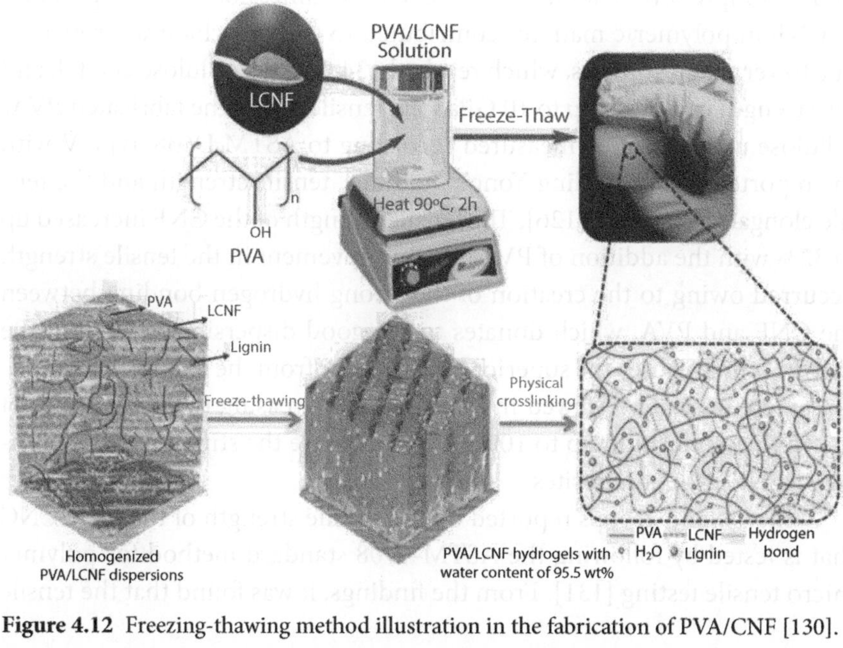

Figure 4.12 Freezing-thawing method illustration in the fabrication of PVA/CNF [130].

double the tensile strength modulus of PVA and with optically transparent composites [127]. Poyraz *et al.* fabricated PVA/CNF, which was extracted from *Eucalyptus camaldulensis* via casting process found out that PVA/CNF has the highest tensile strength compared to pure CNF and PVA/CNF-silica with 141.7±7.8 MPa, 122.4±4.6 MPa and 58.2±10.4 MPa, respectively [129]. The fabrication of PVA/CNF also can be prepared via the freezing-thawing method, as schematically shown in Figure 4.12 [130].

4.3 Properties and Characterizations Techniques

4.3.1 Tensile Characterizations

Tensile strength is indicated as the maximum stress experienced by the composites during the testing. In certain applications, especially in food packaging material, the mechanical properties which cover the tensile strength of the material are very important in order to find out if they could have sufficient mechanical strength during the transportation, handling and storage for packaged food. The right materials of the cellulose with the highly oriented crystalline structure increase the stiffness and the strong in tension but are simultaneously sensitive toward kink band formation under compressive loading [33]. The successfulness of the incorporation of CNF in polymeric matrices contributed to their mechanical properties due to very high stiffness, which reached 134 GPa for cellulose crystal, and the strength can reach up to 10 GPa. The tensile test of the fabricated PVA/cellulose usually can be measured according to ASTM D638, type V with the reported data including Yong's modulus, tensile strength and the tensile elongation at break [126]. The tensile strength of the CNF increased up to 32% with the addition of PVA. The improvement in the tensile strength occurred owing to the creation of the strong hydrogen bonding between the CNF and PVA, which donates to the good dispersion of CNF in the matrix, which causes the superior load transfer from the matrix to the reinforcement, as can be viewed in Figure 4.13, increasing the percentages of the CNF composition up to 10% able to increase the stiffness of the fabricated CNF/PVA composites.

Cameron and groups reported on the tensile strength of the PVA/CNC that is tested by following the ASTM-1708 standard method for polymer micro tensile testing [131]. From the findings, it was found that the tensile

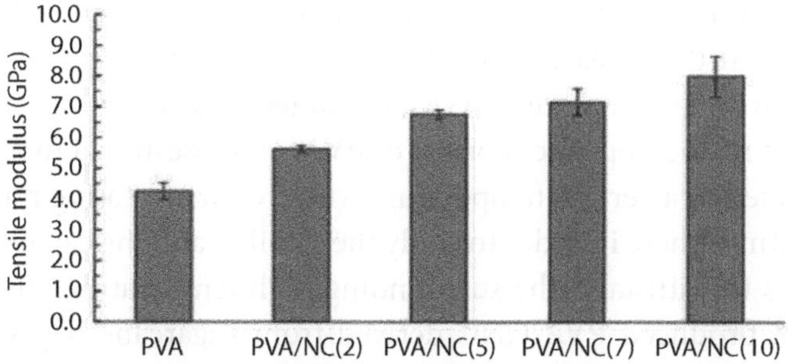

Figure 4.13 Tensile modulus of PVA/CNF composites [126].

strength of the PVA/CNC higher compared to neat PVA with 130±16 MPa and 112±15 MPa, respectively. The strong bonding created between the CNC/PVA increased the binding between them, thus creating a high strength network. In another approach, the PVA reinforced with cellulose microfibers (CMF) from *Jatropha Curcus* L. seed shell improved the tensile strength from 20.69 MPa to 29.32 MPa and Young's modulus from 2.88 GPa to 16.241 Gpa [132]. Better tensile strength is believed contributed by the effective filler dispersion and adhesion between the filler and the matrix. Next, the addition of the CNC in PVA composites is able to increase the tensile strength from 77 MPa to 109 Mpa [133].

4.3.2 Thermal Characterizations

Thermal management has become a decisive concern in the area of electronic devices owing to the surging power densities and growing dependency on polymeric packaging material. The typical thermal conductivity value for the amorphous polymer is very low, which is typically in the range of 0.1 $Wm^{-1}K^{-1}$ to 0.3 $Wm^{-1}K^{-1}$ [134]. This situation will generate a foremost bottleneck in terms of thermal management, which lead to overheating and in overall, will permanently damage the device. In addition, 55% of the electronic equipment failure cases are reported happened due to overheating action. The thermal conductivity can be increased by using an inorganic filler such as boron nitride, metal nitride, metal oxide or clay. However, the addition of these materials may sacrifice other important polymer properties such as processability or electronic insulation. Thus, cellulose-based material that possesses low coefficient thermal expansion and high thermal strength can be potentially binding with PVA as it is a waterborne polymer that can form relatively strong and stable hydrogen bonds. In the research area, thermal analysis can be analyzed via a Thermogravimetric analyzer (TGA) and differential scanning calorimetry (DSC). The nanocomposites consist of PVA/cellulose-based materials that will be heated to a certain temperature with a certain heating rate under a nitrogen atmosphere in order to study the stability and the capability of the composites to withstand the surrounding high temperature. The thermal analysis of produced PVA/CNC derived from sugarcane bagasse is analyzed via TGA by heating the samples from 30°C to 700°C at the heating rate of 10°C/min [135]. It was found that the reinforced PVA/CNC had higher thermal stability compared to pure PVA. A similar finding has been reported previously [8, 117, 136].

4.3.3 X-Ray Diffraction

X-ray diffraction (XRD) technique is a common method to study the crystal structure and atomic spacing. The fundamental in XRD analysis is based on the constructive interference of monochromatic X-rays and crystalline samples. These X-rays are produced by cathode ray tube, filtered to produce monochromatic radiation, collimated to focus, and focussed toward the samples. By scanning the sample through a range of 2θ angles, all the possible diffraction of the lattice could be attained because of the aimless orientation of the powdered materials.

The XRD pattern of PVA/BC composites is displayed in Figure 4.14. Generally, PVA/BC demonstrated the peak of the BC at 14.7°, 16.8°, and 22.6° in the crystal region. Meanwhile, PVA shows the peak at 19.5°, 23.7°, and 27.8° [74]. As the concentration of the BC increased, the intensity of the peak increased and reduced the intensity of PVA. This can be related to the BC content that alters the cross-linking density of the network and disturbs the degree of the crystallinity of the PVA. The crystallinity of the PVA/CNC derived from sugarcane bagasse displayed the crystalline cellulose I pattern with major characteristic peaks assigned at 16°, 22.5° and 34.5°, which corresponding to the diffraction plane at 110, 200, and 004, respectively. Furthermore, the intensity increase of the peak coincided with the increased composition of the CNC content. The addition of the CNC into PVA promotes crystalline structural changes in the PVA network [19].

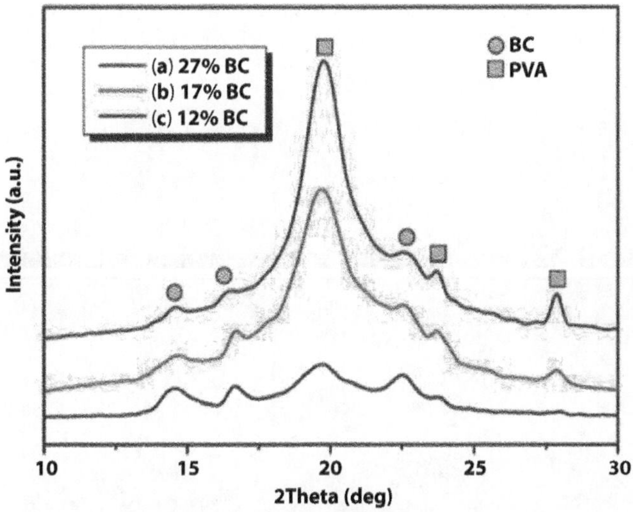

Figure 4.14 XRD pattern of PVA/BC at different concentrations of BC content [74].

4.3.4 Morphological Characterizations

The morphologies of the prepared PVA/cellulose-based materials can be characterized by using various kinds of techniques, including scanning electron microscopy (SEM), transmission electron microscopy (TEM), and field emission scanning electron microscopy (FESEM). The analysis related to the morphologies of the fabricated PVA/cellulose-based materials gave a fully understanding of the physical changes that occurred during the modification process. Different percentage of CNC sourced from sugarcane bagasse incorporated with PVC shows significant observation [19]. The PVA/CNC with the different weight percent of CNC varies from 2-10 wt% affect the porous structure with irregular pores and interconnected pore structure as channels. Clearly, as illustrated in Figure 4.15, the structure of the PVA/CNC porous nanocomposites scaffold showed round or oval pores with an approximate pore size range of 37-164 μm. Increased loading of CNC directly increased the pore size of the PVA/CNC. A similar observation was shown in the fabrication of PVA/CNF, which is derived from the ginger rhizomes (GNF) [137]. The reinforcement of the different ratios of GNF (1 wt% to 7 wt%) with PVA produced uniform and unbroken dispersion of the GNF in the PVA matrix, which will improve

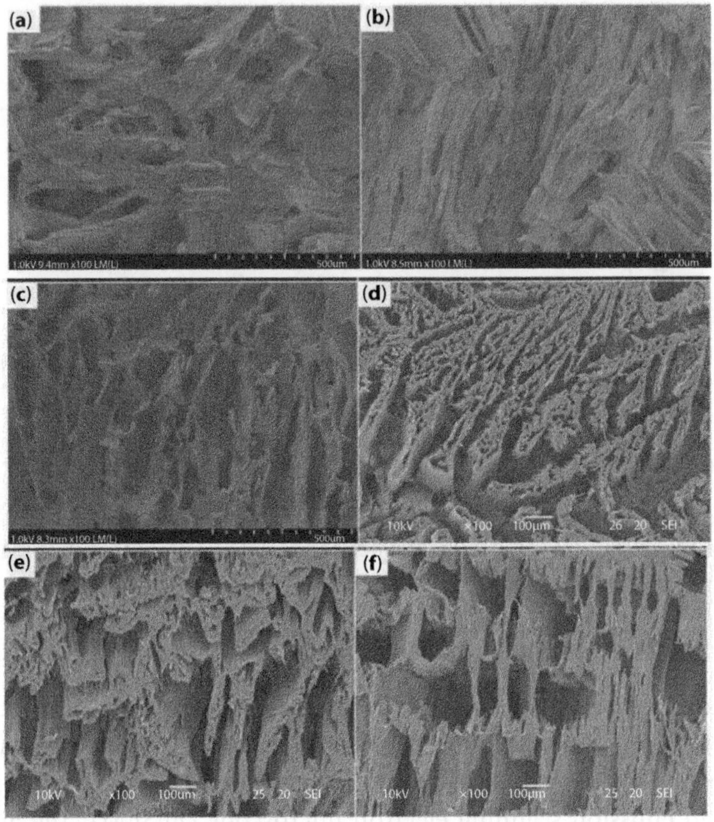

Figure 4.15 SEM images of cross scaffold cross sections at 100 x magnification of (a) PVA; (b) PVA/CNC2; (c) PVA/CNC4; (d) PVA/CNC6; (e) PVA/CNC8 and (f) PVA/CNC10 [19].

the mechanical properties of the nanocomposites [138, 139]. The casted membrane of PVA with commercialized CNC at different concentrations of 0.5 wt% to 2 wt% was created to be applied in CO_2/CH_4 gas separation [25]. The membrane morphology was analyzed by using SEM shows that the membrane contains 1 wt% of CNC display smooth and zero defects. It was observed that the addition of the CNC increased the thickness of the selective layer on the casted PVA/CNC membrane as the concentration of the CNC increased, which is directly increased the viscosity of the casting suspension.

Other than that, the development of biocomposites potentially applied in enzymatic treatment for biotechnological exploitability is fabricated via the development of PVA/BC. The delignification of wheat straw based-PVA assisted with BC improved the mechanical and wettability features [140]. Noticeably, a deposition of BC is forming randomly at the cellulosic microfibrils in Figure 4.16 (c) compared to pristine PVA in Figure 4.16 (a). In addition, the bacterially modified wheat straw (PVA/BMWS) showed more fibrous, indicating the increased amount of cellulosic content, which is directly boost the mechanical properties of the composites. Formerly, Abral and groups fabricated PVA/BC pellicles blended with cassava starch blends [75]. Fabrication of PVA/CNF derived from *Eucalyptus camaldulensis* was observed to have much better dispersion compared to PVA/CNF-silica [129]. As a review in SEM images in Figure 4.17, a better dispersion shows good adhesion and homogeneous

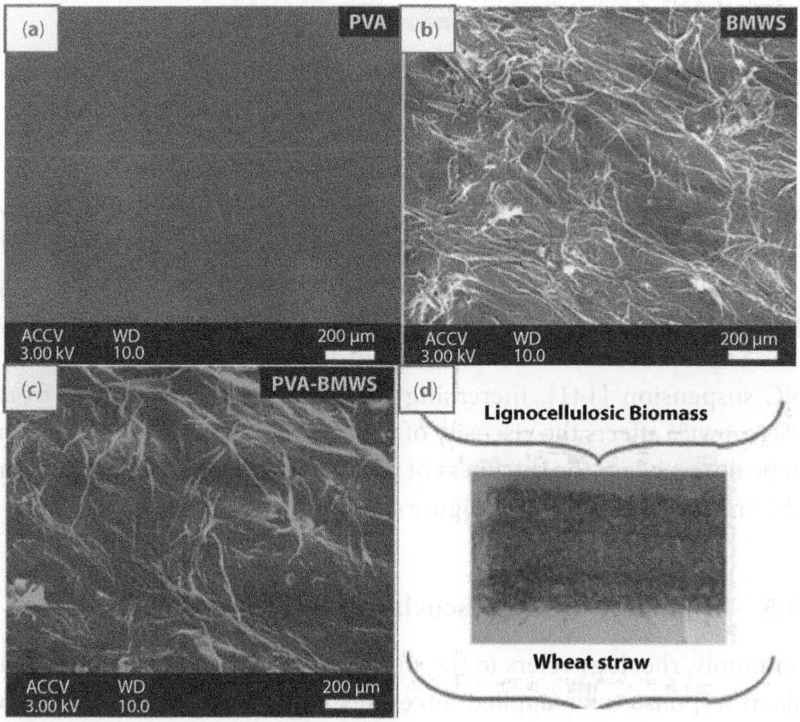

Figure 4.16 Surface morphology of (a) PVA; (b) BMWS; (c) PVA/BMWS; (d) lignocellulosic biomass of wheat straw [140].

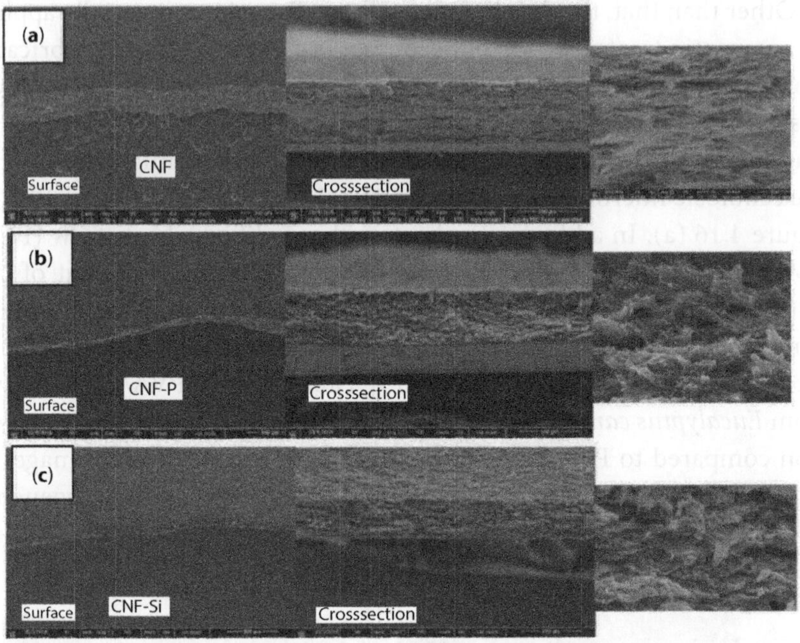

Figure 4.17 Surface and cross-section images via SEM analysis of fabricated (a) pristine CNF; (b) PVA/CNF, and (c) PVA/CNF-silica [129].

distribution of the CNF and PVA. Similar morphology being observed by Gruneberger and groups, which proved the better homogenization between the PVA/CNF than PVA/CNF-silica [114]. Specifically, in gas separation application, the fabricated PVA/CNC nanocomposites membrane is prepared with the variation of the different concentrations of CNC suspension [141]. Increasing the concentrations of CNC from 0 wt% to 4 wt% affects the viscosity of the casting suspension. The outcome can be observed on the thickness of the selective layer that can be seen in SEM images, as exhibited in Figure 4.18.

4.3.5 Rheological and Viscoelastic Characterizations

Commonly, rheology refers to the study of flow and deformation of materials in response to an applied force. In specific, theological behaviour is one of the properties of the hydrogel or aerogel of the composites since

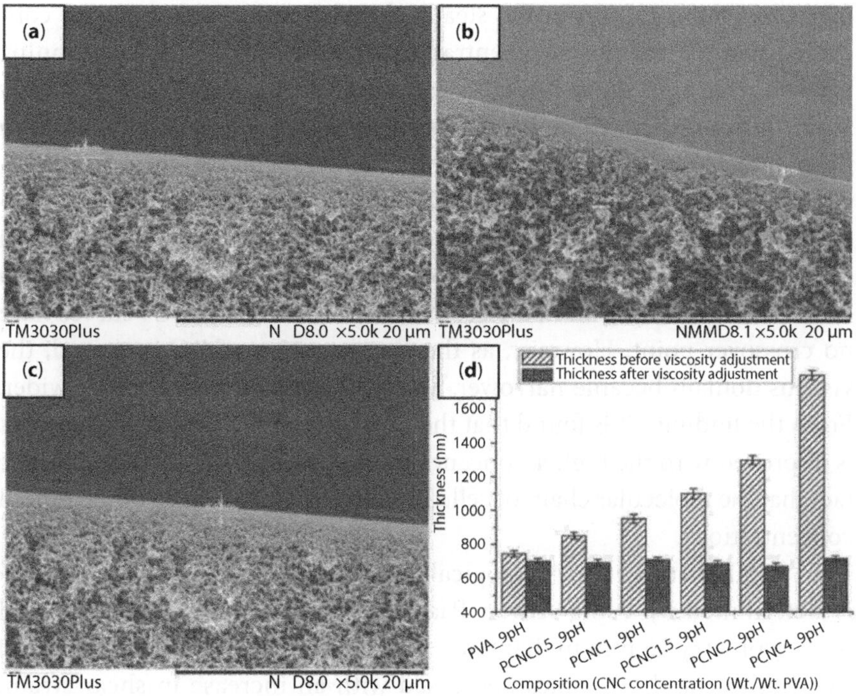

Figure 4.18 Morphology studies of (a) cross-section of PVA membrane; (b) 2 wt% of CNC/PVA membrane before viscosity adjustment; (c) 2 wt% of CNC/PVA membrane after viscosity adjustment and (d) selective layer compared before and after viscosity adjustment [141].

it involved the freezing and thawing cycle that allowed the researchers to study the network formation of the gels [142]. The measurement of the rheological can be measured by using a rheometer equipped with a Peltier device for rigorous temperature control. The viscoelastic parameter of the hydrogel by a small-amplitude oscillatory shear test with certain ranging of oscillation frequency and shear strain (y) at which linear viscoelasticity is attained. The storage (*G'*) and loss (*G"*) are measured as a function of the oscillation frequency (ω̇). *G'* linked with the elasticity or the energy stored in the material during the deformation. Meanwhile, *G"* gives the information about the viscous character or the dissipated energy in the environment [109].

The rheological properties of the RC derived from the dissolution of cotton pulp in ionic liquid can be studied by steady and oscillatory shear measurement [143]. From the steady shear viscosity test, it can be concluded that the cellulose concentration shows a common shear-thinning behaviour at high shear rates. As can be scrutinized in Figure 4.19, the shear-thinning behaviour of the solution is the consequence of both shear-induced reductions in the number of entanglements and the desolvation of macromolecules by the high shear rate [144]. Meanwhile, the viscoelasticity of the RC is studied by analyzed the *G'* and *G"* as a function of frequency for different cellulose concentrations. At a low concentration of RC, the solution form in a viscous fluid, and the *G'* and *G"* curves have no crossover point. However, as the concentration of RC increased, the viscous domain became narrower, and the elastic domain became wider. From the findings, it is found that the capability of the temporary network is improved with the higher concentration of RC that can be related to the fact that the molecular chain of cellulose is more close to each other at high concentrations.

In another study, the rheological behaviour of hydrogel of PVA/CNF is determined by using Anton Paar MCR302 rheometer with parallel plate geometry 25 mm in diameter [130]. From the data, it is discovered that both *G'* and *G"* values decreased with an increase in shear strain. As assumed, due to the entanglement and cross-linking of CNF in PVA hydrogel demonstrated higher *G'* and *G"* particularly when the loading concentration of CNF increased up to 2 wt%. For instance, *G'* value of PVA/CNF with 2 wt% is 7725 Pa compared to 0.25 wt% CNF reinforced

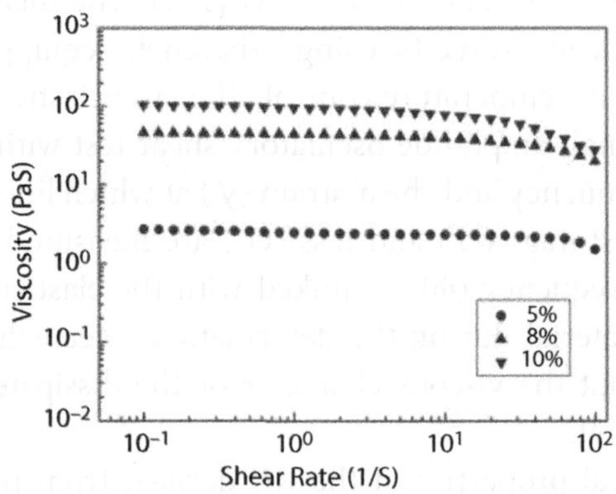

Figure 4.19 Viscosity-shear rate of RC in ionic liquid with different concentrations [143].

in PVA hydrogel with the value of 492 Pa. In another approach, the rheological characterization of the reinforced cellulose nanoparticles, CN contains CNC and CNF in PVA hydrogel, is measured by using a stress-controlled rheometer and steady-state viscosity is calculated in a wide shear rate range from 0.1 s^{-1} to 1000 s^{-1} [145]. The steady-state viscosities as a function of the shear rate for CN suspensions are presented in Figure 4.20. Overall, the suspension having look-alike shear thinning behaviour which is characterized by the decrease of viscosity with increasing shear rate. A higher concentration of CN is parallel to the viscosity along the whole curve on account of the increased concentration of CNF aid the formation of a more rigid and entangled network among fibrils, as illustrated in Figure 4.21.

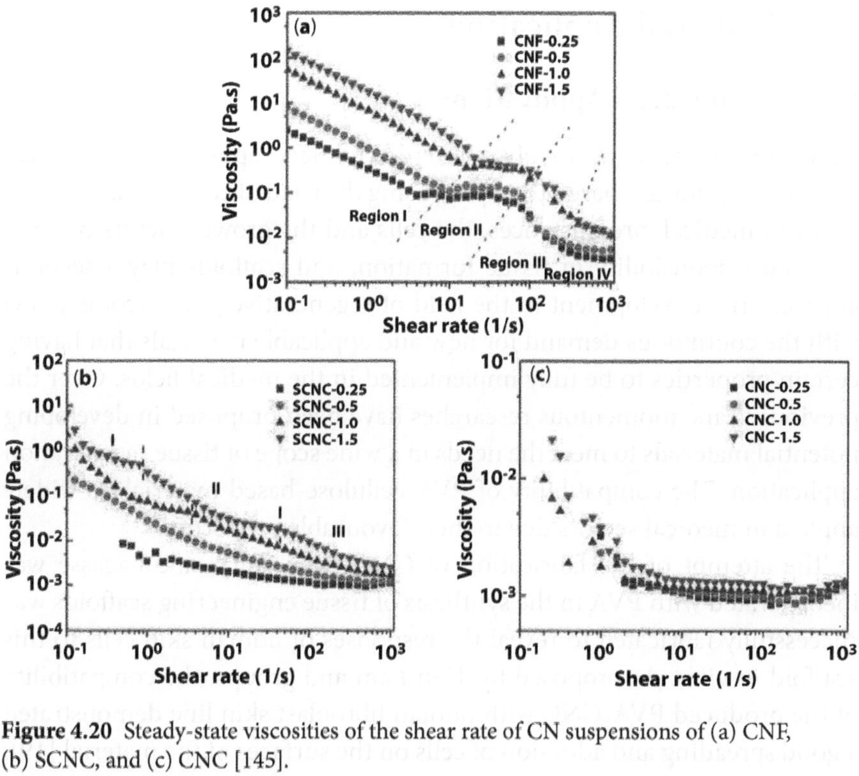

Figure 4.20 Steady-state viscosities of the shear rate of CN suspensions of (a) CNF, (b) SCNC, and (c) CNC [145].

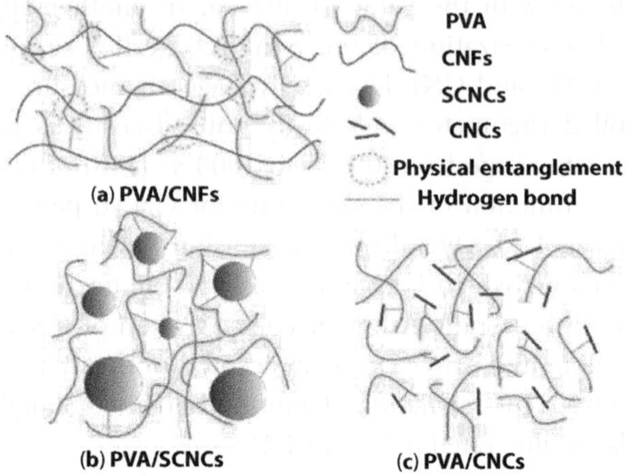

Figure 4.21 Schematic illustration of the network of (a) PVA/CNF suspension, (b) PVA/SCNC, and (c) PVA/CNC suspension [145].

4.4 Potential Applications

4.4.1 Biomedical Applications

Regenerative medicine involves interdisciplinary approaches for tissue engineering and cell-based therapies, using the triad of cells, signals or scaffolds. In medical circumstances, the cells and the growth factors become the vital responsibility in tissue formation, and scaffolds play a secondary role. The development in the field of regenerative goes incomparable with the continuous demand for new and applicable materials that having certain properties to be fully implemented in the medical fields. Over the previous years, momentous researches have been proposed in developing potential materials to meet the needs in a wide scope of tissue regeneration application. The compatibility of PVA/cellulose-based materials is widely applied in medical sectors due to their favourable properties.

The attempt of the fabrication of CNC from sugarcane bagasse was incorporated with PVA in the synthesis of tissue engineering scaffolds was successfully fabricated to reveal the responses of human skin cells to this scaffold *in vitro*. As proposed by Tien Lam and groups, the compatibility of the produced PVA/CNC with human fibroblast skin line demonstrated a good spreading and adhesion of cells on the surfaces of the material [19]. As illustrated in Figure 4.22, on the seventh day of culture, the cell rapidly

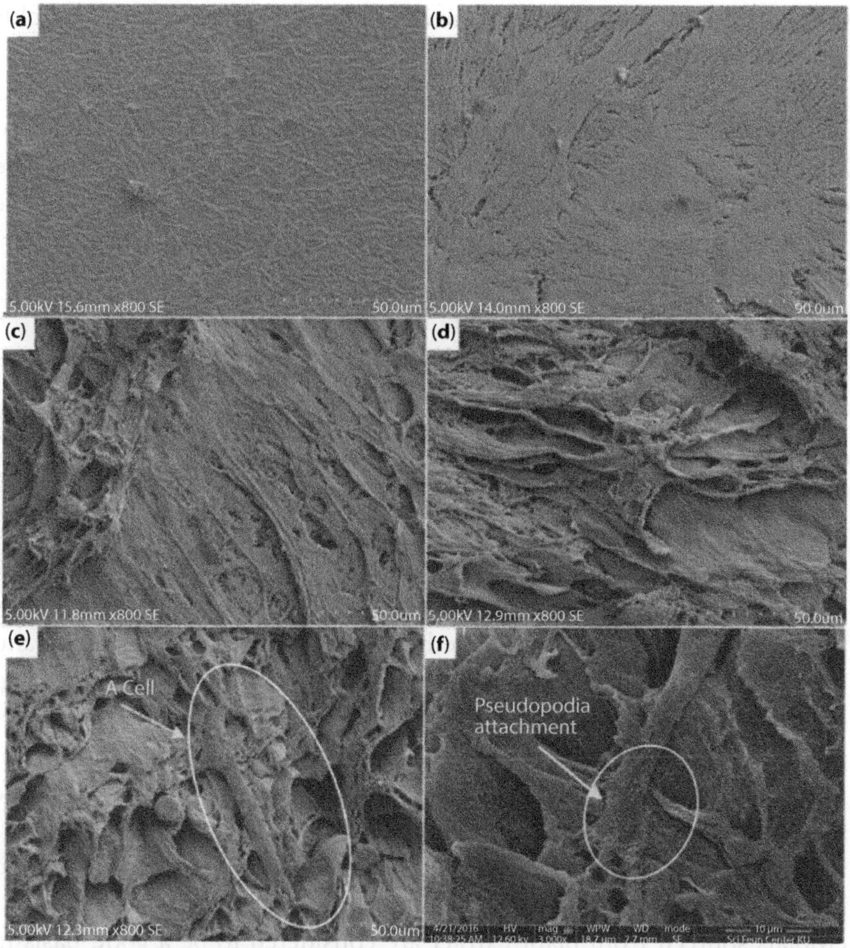

Figure 4.22 SEM images of the surface section of the (a) positive control; (b) negative control; (c) PVA; (d-f) PVA/CNC after 7 days of incubation [19].

grew with the culture time, well spread and completely covered the surfaces. The PVA/CNC shows a uniform, porous morphology without any shrinkage after the cell growth. The SEM images also exemplified that zero interior spaces between the cell and the surface material, which proved the unblemished adhesion of the scaffold and the cells. The findings attested that PVA/CNC had not revealed any toxicity for human fibroblast skin cell regeneration. In another publication, the result of the antibacterial rate against Gram-negative bacteria *E. coli* and *S. aureus* by chitosan/PVA/

CNC over 80% and 76%, respectively, indicated the fabricated scaffolds having good bactericidal activity [18].

Other than a scaffold for tissue engineering, the application of PVA/BC hydrogel composite shows some potential use as artificial cornea biomaterial [74]. Desirable properties resulted from the PVA/BC composite exhibited high water content, high visible light transmittance, and suitable UV absorbance, and good mechanical strength make it as promising characteristics as an artificial cornea.

4.4.2 Packaging Applications

The nutritional quality of food and its safety is decisive to the evolution of functional packaging, which is a new class of packaging that comprises some purpose for the armour of the food product compared to only supplying fence materials regarding the gases and moisture content. Thus, the existing functional material can be amalgamated into new fabricated new packaging material to impart various purposes in contrast with the pristine packaging. On account of the health concern of the consumers and the environmental crisis, the recent fact-finding in functional packaging has meticulously studied the use of natural functional materials in biodegradable polymeric functional materials. The antimicrobial properties in packaging are the most crucial characteristic since it will determine the shelf life of the food product. In short, the duration of the product will last longer with the good antimicrobial packaging properties owing to good inhibition of the growth of microorganisms in food. The natural antimicrobials sourced from plant extract are considered to be a good basis to assure food safety.

The fabrication of composite film containing CNC derived from red seaweed and basil leaves extract (BE) were incorporated in PVA was successfully fabricated as food packaging that having antimicrobial and antioxidant properties, respectively [146]. As mentioned by Singh and co-workers, BE, which is a culinary herb, consists of phytochemicals constituents that are commonly known for its capability to obstruct the growth of pathogens. Several researchers also reported the effectiveness of the extract composition from the basil that being able to slow down the production of gram-positive and gram-negative bacteria [147–149]. The antimicrobial activity of the BE and PVA/CNC film having different concentrations of basil leaves, which are 0, 5, 10, and 20% toward *L. monocytogenes* and

B. Cereus bacterial, is presented in Figure 4.23. From the observation, the pure BE shows a higher inhibition zone compared to the prepared film with BE. The prepared bio-nanocomposites films containing 5 and 10% BE did not show signs of inhibition region for *L. monocytogenes* and *B. Cereus.* However, PVA/CNC film containing 20% of BE displayed a spotless area for *B. Cereus.* In contrast, there is no inhibition zone for *L. Monocytogenes,* which verified that the film owing bacteriostatic effect toward for *L. monocytogenes* and *B. Cereus.*

In another approach, the agro-industrial waste obtained bio-constituent from ginger is successfully isolated to produce 100 to 200 nm of CNF. The isolation of the CNF sourced from ginger having good antibacterial activity against *Bacillus cereus, Escherichia coli, Staphylococcus aureus* and

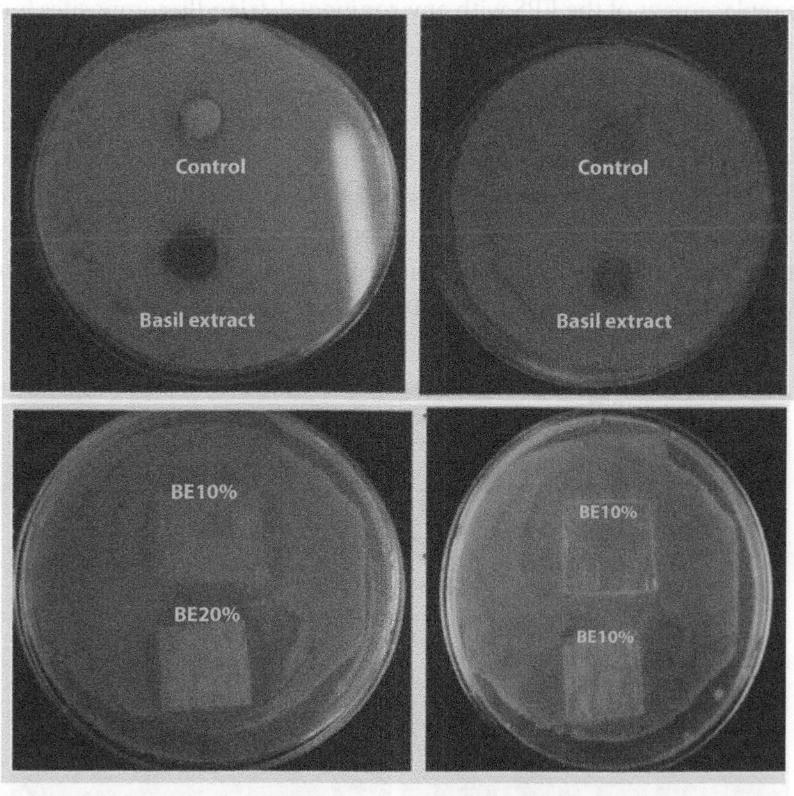

Figure 4.23 Antimicrobial activity of BE (top) and PVA/CNC/BE film with 10% and 20% BE (below) against *B. Cereus* (left) and *L. monocytogenes* (right) [146].

Salmonella typhimirium [137]. The ability of the CNF sourced from ginger to possess good activities against microbes is improved by reinforcement of chitosan (CS) and PVA. Both the CS and PVA is believed to be able to boost the mechanical and properties of these fabricated bionanocomposites. In addition, the packaging materials normally demand certain barrier properties as well as mechanical resistance. Thus, the implementation of the PVA in cellulose-based biocomposites highly demand in packaging materials as PVA itself possess certain resistance to grease, oils and solvents. The study that is conducted by and Kalnins proved that the PVA/cellulose fiber composites fulfil the packaging obligation [36, 150]. In addition, the foam plate that is extracted from potato starch and corn fibers reinforced with PVA is believed to be able to substitute single-use packaging materials made up of expanded polystyrene (EPS) due to its biodegradability [151]. As we know, the plate which comes from the EPS is difficult to be decomposed. Thus, the replacement of the EPS with environmental-friendly composites is one of the approaches in order to reduce the consumption of non-renewable and non-degradable materials. In another work, the fabricated corn starch/PVA/CF derived from *Grewia Optiva* fiber is fabricated and proposed to have antibacterial activity against *Staphylococcus aureus* and *Escherichia coli*. The observation from the Figure 4.24 shows that the blended film can be used as an exceptional material for food packaging [35]. The film shows antibacterial activity against Gram-positive (*Staphylococcus aureus*) and Gram negative (*Escherichia coli*). Inhibitory zone against *Staphylococcus*

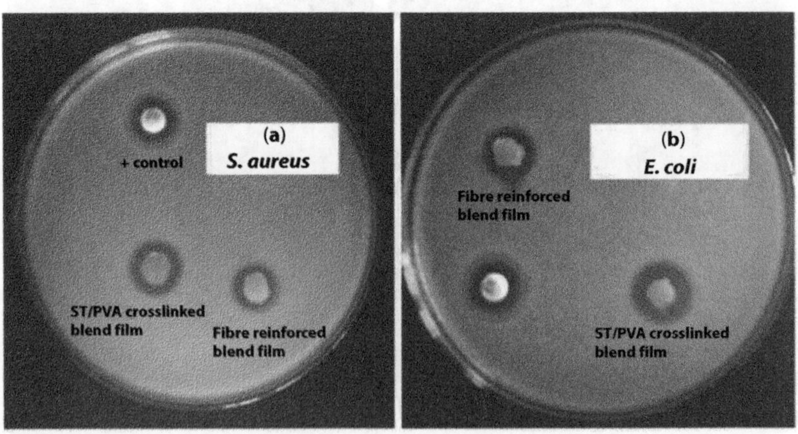

Figure 4.24 Antibacterial effect of corn starch/PVA and corn starch/PVA/CF against *Staphylococcus aureus* and *Escherichia coli* [35].

aureus and *Escherichia coli* is measured to be 1.5 and 1.2 cm, 1.7 cm and 1.5 cm for corn starch/PVA and corn starch/PVA/CF, respectively. The PVA/CNC/chitosan bionanocomposites film also exhibited good antifungal and antibacterial activity toward *Colletotrichum gloeosporioides* and *Lasiodiplodia theobromae* [121]. The inhibition toward the bacteria proved that PVA/CNC/chitosan bionanocomposites film has a great potential to be utilized in the food packaging application.

4.4.3 Heavy Metal Applications

By-product industrial field has become a serious environmental problem worldwide. The composition of the effluent contains series of contaminants, including toxic heavy metal ions, inorganic anions, organic compounds such as dyes, phenols, pesticides and other sustained organic pollutants [152–155]. The consequences that come from the hazardous pollutants majorly have a negative effect on human beings and the aquatic lives systems. Thus, it is vital for the researchers to cover a new alternative with good approaches to minimize the problems effectively. The heavy metals application covered many methods including adsorption [152, 156, 157], membrane technology [158–160], flocculation [155, 161, 162], filtration [163–165], ion exchanges [166, 167] and etc. According to many research studies, the adsorption technique has a high potential for the removal of heavy metals from wastewater [156, 168–173]. Basically, adsorption refers to the mass transfer process by means of transferring ion particles of heavy metals from the liquid phase in wastewater to the surface of the solid by adsorbents.

Year by year, the publication which focuses on the materials for the removal of heavy metals kept increasing, showing that these problematic issues need to give special attention. There are many types of nanocellulose-based adsorbents commonly used as a medium to separate heavy metals from wastewater. Cellulose-based biocomposites are widely used as adsorbents since they provide higher adsorption capacities and offer excellent binding affinities toward heavy metal ions, compared with the macroscale structure of cellulosic materials. Interestingly, the capability of the PVA-cellulose based biocomposites shows some favourable properties in order to minimize the high level of the metal ions in the wastewater and real industrial fields. The fabrication of PVA/CA with the addition of the magnetic properties of Fe_3O_4 was done by Nouri and groups [49].

Table 4.1 Application of PVA-cellulose based biocomposites and bionanocomposites in heavy metal removal.

Type of adsorbents	Targeted heavy metals	Adsorption capacity (mg/g)/percentage removal (%)	References
PVA/CNC	Pb(II)	471.55 mg/g	[174]
PVA/CNC hydrogel	Ag(I)	8.4 mg/g	[175]
PVA/CNF	Cu(II) Pb(II)	30.0 mg/g 131.5 mg/g	[176]
PVA/CNC	Cu(II) Pb(II)	484.0 mg/g 323.4 mg/g	[177]
PVA/CNF	Hg(II) Pb(II) Cu(II)	157.5 mg/g 22.0 mg/g 110.6 mg/g	[178]

The produced nanocomposite thin film of PVA/CA/Fe_3O_4 is applied for the adsorption of cadmium, Cd(II) and nickel, Ni(II) metals ions. Table 4.1 summarizes the utilization of PVA-cellulose based biocomposites and bionanocomposites in heavy metal removal via adsorption approaches.

4.4.4 Gas Separation

The world is now experiencing a brisk increase in population, resulting in higher demands on industrial production and energy consumption with every passing day. Globally, fossil fuels play a dominant role in order to meet energy needs. The production and the excessive use of fossil fuels have triggered global warming due to the release of carbon dioxide, CO_2. The emission of CO_2, which comes from the combustion of fossil fuels and certain industries, has increased up to 61% from 1990 till 2013 [141]. Thus, the amount of CO_2 released into the atmosphere kept increasing year by year and is expected to drastically discharge in the future. In addition, CO_2 is predominating anthropogenic greenhouse gas. The release of CO_2 into the environment is responsible for 77% of human contribution toward the greenhouse effect. This troubling situation has driven the public interest and attention toward the utilization of renewable energy sources. Thus, the fabrication of PVA incorporated with cellulose-based nanomaterials regained trusted materials to be applied in the gas separation application.

Recent research has been done by Torstensen and colleagues in applying the CNC nanocomposites in PVA to form film membranes for CO_2 separation from flue gas [27]. From the finding, PVA/CNC possesses better performance with CO_2 permeance reached 127.8±5.5 GPU with CO_2/N_2 separation factor (39±0.4). Contradict, PVA composite membrane having lower CO_2 permeance (105.5±1.9 GPU) with CO_2/N_2 separation factor of 36±0.5. The introduction of CNC to PVA for CO_2 improves the redistribution of water molecules and $CO_2/HCO_3^-/CO_3^{2-}$ in the nanocomposites to the PVA/CNC interphase. Thus, the creation of the interphase generated a favourable path for transport in which the molecules can easily pass through due to higher gas/ion solubility and less complex ways. The proposed transport mode of the gas through the PVA/CNC nanocomposites is demonstrated in Figure 4.25. In other approaches, the fabricated PVA/CNC shows higher selectivity up to 39 and achieved permeance with 0.27 m^3 (STP)/(m^2 bar h) for CO_2/CH_4 separation [141]. In a similar application, the nanocomposites membranes of PVA/CNC proved that the addition of CNC affect the degree of swelling, crystallinity and thickness of the resulting membrane, with the permeation testing showed that the permeance and the selectivity for CO_2 increased with the addition of CNC [25]. The maximum permeance obtained is 0.29m^3 (STP)/(m^2 bar h), and the selectivity of CO_2 over CH_4 is 43.

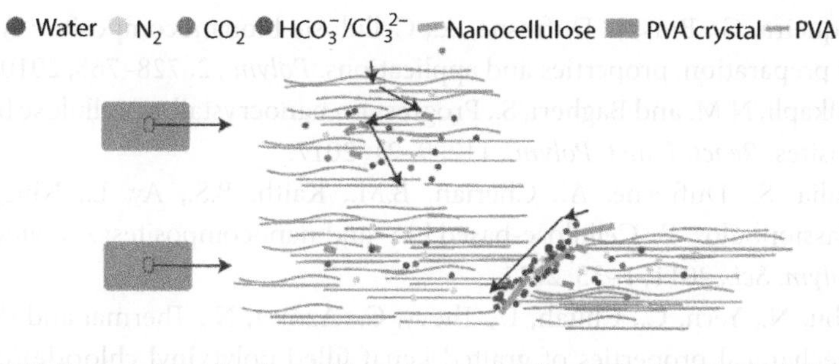

Figure 4.25 Proposed mechanism of the transport mode of CO_2 gases in PVA/CNC nanocomposites membrane [27].

4.5 Conclusion

Cellulose is the most abundant sustainable resource on the earth, which has been used for thousands of years by mankind. In the 20th century, petrochemical resources dominated the primary resources in the industry, which was the main material source and prominent our life quality. However, the growing concern regarding the use of petroleum-based composites due to their non-biodegradability and non-renewability. Thus, coming to the 21st century, biomass materials have become noticeable materials owing to the increasing terrible environmental crisis. Specifically, biopolymer-based composites and blends can suitably address the challenge raised by their synthetic counterparts. Cellulose is the most abundant natural polymer with promising characteristics as composites materials. Cellulose-based materials that come from many types of sources, preparation, and properties produced different classes of cellulose, including CNC, RC, BC, hydrogel-types, CNF, and CF, which has a different capability to be applied for different applications. The cellulose-based incorporated with PVA increase the chances for the produced nanocomposites to have excellent properties in terms of mechanical, stiffness and surface reactivity. Moreover, the PVA/cellulose-based biocomposites and bionanocomposites are able to widen the application of the materials, which covered the biomedical application, gas separation, packaging application, and many more challengeable fields in the future.

References

1. Siqueira, G., Bras, J., Dufresne, A., Cellulosic bionanocomposites: A review of preparation, properties and applications. *Polym.*, 2, 728–765, 2010.
2. Julkapli, N.M. and Bagheri, S., Progress on nanocrystalline cellulose biocomposites. *React. Funct. Polym.*, 112, 9–21, 2017.
3. Kalia, S., Dufresne, A., Cherian, B.M., Kaith, B.S., Av, L., Njuguna, J., Nassiopoulos, E., Cellulose-based bio- and nanocomposites: A review. *Int. J. Polym. Sci.*, 2011, 1–35, 2011.
4. Abu, N., Yern, C., Chuah, L., Thevy, C., Azowa, N., Thermal and dynamic mechanical properties of grafted kenaf filled poly(vinyl chloride)/ethylene vinyl acetate composites. *J. Mater.*, 65, 204–211, 2015.

5. Tan, B.K., Ching, Y.C., Poh, S.C., Abdullah, L.C., A review of natural fiber reinforced poly(vinyl alcohol) based composites: Application and opportunity. *Polym.*, 7, 2205–2222, 2015.
6. Cazón, P., Vázquez, M., Velazquez, G., Composite films of regenerate cellulose with chitosan and polyvinyl alcohol: Evaluation of water adsorption, mechanical and optical properties. *Int. J. Biol. Macromol.*, 117, 235–246, 2018.
7. Jana, S., Saikia, A., Purkait, M.K., Mohanty, K., Chitosan based ceramic ultrafiltration membrane: Preparation, characterization and application to remove Hg(II) and As(III) using polymer enhanced ultrafiltration. *Chem. Eng. J.*, 170, 209–219, 2011.
8. Bilal, M., Niazi, K., Jahan, Z., So, S., Weiby, Ø., Mechanical, thermal and swelling properties of phosphorylated nanocellulose fibrils/PVA nanocomposite membranes. *Carbohydr. Polym.*, 177, 258–268, 2017.
9. Choo, K., Ching, Y.C., Chuah, C.H., Julai, S., Liou, N., Preparation and characterization of polyvinyl alcohol-chitosan composite films reinforced with cellulose nanofiber. *Materials*, 9, 1–16, 2016.
10. Luo, L., Cai, W., Zhou, J., Li, Y., Facile synthesis of boehmite/PVA composite membrane with enhanced adsorption performance towards Cr(VI). *J. Hazard. Mater.*, 318, 452–459, 2016.
11. Zhang, L., Zhang, R., Yang, Q., Qi, H., Zhou, J., Zhang, L., Preparation and properties of cellulose/poly(vinyl alcohol) blend films based on dissolution in a non-toxic solvent system preparation and properties of cellulose/poly (vinyl alcohol) blend films based on dissolution in a non-toxic solvent system. *J. Biobased Mater. Bioenergy*, 3, 199–204, 2009.
12. Zelikin, A.N., Poly(vinyl alcohol) physical hydrogels: Matrix-mediated drug delivery using spontaneously eroding substrate. *J. Phys. Chem.*, 120, 5916–5926, 2016.
13. Muppalaneni, S. and Omidian, H., Polyvinyl alcohol in medicine and pharmacy: A perspective. *J. Drug Delivery*, 2, 1–5, 2013.
14. Tak, H., Yun, Y., Lee, C., Yoon, S., Sulindac imprinted mungbean starch/PVA biomaterial films as a transdermal drug delivery patch. *Carbohydr. Polym.*, 208, 261–268, 2019.
15. Vashisth, P., Raghuwanshi, N., Srivastava, A.K., Singh, H., Nagar, H., Pruthi, V., Ofloxacin loaded gellan/PVA nano fibers - Synthesis, characterization and evaluation of their gastroretentive/mucoadhesive drug delivery potential. *Mater. Sci. Eng. C*, 71, 611–619, 2017.
16. Vrana, N.E., Liu, Y., Mcguinness, G.B., Cahill, P.A., Characterization of poly (vinyl alcohol)/chitosan hydrogels as vascular tissue engineering scaffolds. *Macromol. Symp.*, 269, 106–110, 2008.

17. Kumar, A. and Han, S.S., PVA-based hydrogels for tissue engineering: A review. *Int. J. Polym. Mater. Polym. Biomater.*, 66, 159–182, 2017.
18. Kanimozhi, K., Basha, S.K., Kumari, V.S., Processing and characterization of chitosan/PVA and methylcellulose porous scaffolds for tissue engineering. *Mater. Sci. Eng. C*, 61, 484–491, 2016.
19. Tien, N., Chollakup, R., Smitthipong, W., Nimchua, T., Utilizing cellulose from sugarcane bagasse mixed with poly(vinyl alcohol) for tissue engineering scaffold fabrication. *Ind. Crops Prod.*, 100, 183–197, 2017.
20. Zhou, Y., Zhao, S., Zhang, C., Liang, K., Li, J., Yang, H., Photopolymerized maleilated chitosan/thiol-terminated poly (vinyl alcohol) hydrogels as potential tissue engineering scaffolds. *Carbohydr. Polym.*, 184, 383–389, 2018.
21. Hanani, F., Shahitha, F., Hussain, J., Harun, W.S.W., Yusoff, M.M., Highly porous of hydroxyethyl cellulose biocomposite scaffolds for tissue engineering. *Int. J. Biol. Macromol.*, 122, 562–571, 2019.
22. Gan, I. and Chow, W.S., Antimicrobial poly (lactic acid)/cellulose bionanocomposite for food packaging application: A review. *Food Packag. Shelf Life*, 17, 150–161, 2018.
23. Salman, M., Bilal, M., Niazi, K., Jahan, Z., Ahmad, T., Preparation and characterization of PVA/nanocellulose/Ag nanocomposite films for antimicrobial food packaging. *Carbohydr. Polym.*, 184, 453–464, 2018.
24. Luzi, F., Fortunati, E., Giovanale, G., Mazzaglia, A., Torre, L., Mariano, G., Cellulose nanocrystals from Actinidia deliciosa pruning residues combined with carvacrol in PVA CH films with antioxidant/antimicrobial properties for packaging applications. *Int. J. Biol. Macromol.*, 104, 43–55, 2017.
25. Jahan, Z., Bilal, M., Niazi, K., Hägg, M., Weiby, Ø., Cellulose nanocrystal/PVA nanocomposite membranes for CO_2/CH_4 separation at high pressure. *J. Membr. Sci.*, 554, 275–281, 2018.
26. Rizwan, M., Islam, A., Sabir, A., Sha, M., Taqi, M., Butt, Z., Ijaz, A., Jamil, T., Peg, P.V.A., Fabrication and performance characterization of novel zinc oxide filled cross-linked PVA/PEG 600 blended membranes for CO_2/N_2 separation. *J. Ind. Eng. Chem.*, 55, 65–73, 2017.
27. Torstensen, J.Ø., Helberg, R.M.L., Deng, L., Gregersen, Ø.W., PVA/nanocellulose nanocomposite membranes for CO_2 separation from flue gas. *Int. J. Greenh. Gas Control*, 81, 93–102, 2019.
28. Huang, J., Lyu, S., Fu, F., Wang, S., RSC advances green preparation of a cellulose nanocrystals/polyvinyl alcohol composite superhydrophobic. *RSC Adv.*, 7, 20152–20159, 2017.
29. Rahimi, M., Kameli, P., Ranjbar, M., Salamati, H., The effect of polyvinyl alcohol (PVA) coating on structural, magnetic properties and spin dynamics of Ni 0.3 Zn 0.7 Fe_2O_4 ferrite nanoparticles. *J. Magn. Magn. Mater.*, 347, 139–145, 2013.

30. Mathew, B., Lopes, A., Ferreira, S., Souza, D., Maria, L., Costa, M., Molina, G., Olyveira, D., Kottaisamy, M., Nagarajan, E.R., Thomas, S., Cellulose nanocomposites with nanofibres isolated from pineapple leaf fibers for medical applications. *Carbohydr. Polym.*, 86, 1790–1798, 2011.
31. Ji, F., Li, C., Tang, B., Xu, J., Lu, G., Liu, P., Preparation of cellulose acetate/zeolite composite fiber and its adsorption behavior for heavy metal ions in aqueous solution. *Chem. Eng. J.*, 209, 325–333, 2012.
32. Abdul Khalil, H.P.S., Bhat, A.H., Ireana Yusra, A.F., Green composites from sustainable cellulose nanofibrils: A review. *Carbohydr. Polym.*, 87, 963–979, 2012.
33. Box, P.O. and Wageningen, A.A., Tensile and compressive properties of flax fibres. *J. Mater. Sci.*, 7, 1683–1692, 2002.
34. Niu, Y., Zhang, X., He, X., Zhao, J., Zhang, W., Lu, C., Effective dispersion and crosslinking in PVA/cellulose fiber biocomposites via solid-state mechanochemistry. *Int. J. Biol. Macromol.*, 72, 855–861, 2015.
35. Priya, B., Kumar, V., Pathania, D., Singh, A., Synthesis , characterization and antibacterial activity of biodegradable starch/PVA composite films reinforced with cellulosic fibre. *Carbohydr. Polym.*, 109, 171–179, 2014.
36. Cerpakovska, D. and Kalnins, M., Barrier and sorption properties of porous poly (vinyl alcohol)-cellulose fibre composites. *Prog. Polym. Sci.*, 61, 178–184, 2012.
37. Vinodhini, P.A. and Sudha, P.N., Removal of heavy metal chromium from tannery effluent using ultrafiltration membrane. *Text. Clothing Sustainability*, 2, 2–15, 2016.
38. Kim, K., Ingole, P.G., Yun, S., Choi, W., Kim, J., Lee, H., Water vapor removal using CA/PEG blending materials coated hollow fiber membrane. *J. Chem. Technol. Biotechnol.*, 90, 1117–1123, 2015.
39. Krason, J. and Pietrzak, R., Membranes obtained on the basis of cellulose acetate and their use in removal of metal ions from liquid phase. *Pol. J. Chem. Technol.*, 18, 104–110, 2016.
40. Ahmad, A., Jamshed, F., Riaz, T., Gul, S.-, Waheed, S., Sabir, A., AlAnezi, A.A., Adrees, M., Jamil, T., Self-sterilized composite membranes of cellulose acetate/polyethylene glycol for water desalination. *Carbohydr. Polym.*, 149, 207–216, 2016.
41. Lee, H., Nishino, M., Sohn, D., Soon, J., Control of the morphology of cellulose acetate nanofibers via electrospinning. *Cellulose*, 25, 2829–2837, 2018.
42. Wahab, A., Soo, I., Ni, Q., Cellulose acetate nano fibers embedded with AgNPs anchored TiO_2 nanoparticles for long term excellent antibacterial applications. *Carbohydr. Polym.*, 207, 640–649, 2019.

43. Kamal, T., Ahmad, I., Bahadar, S., Asiri, A.M., Synthesis and catalytic properties of silver nanoparticles supported on porous cellulose acetate sheets and wet-spun fibers. *Carbohydr. Polym.*, 157, 294–302, 2017.
44. Rajan, A., Gnanasekaran, G., Sathishkumar, Y., Soo, Y., Sang, C., Electrospun antibacterial polyurethane – cellulose acetate – zein composite mats for wound dressing. *Carbohydr. Polym.*, 102, 884–892, 2014.
45. Inukai, S., Kurokawa, N., Hotta, A., Annealing and saponification of electrospun cellulose-acetate nano fibers used as reinforcement materials for composites. *Compos. Part. A*, 113, 158–165, 2018.
46. Taylor, P., Yin, J., Fan, H., Zhou, J., Cellulose acetate/poly(vinyl alcohol) and cellulose acetate/crosslinked poly(vinyl alcohol) blend membranes: Preparation, characterization, and antifouling properties. *Desalin. Water Treat.*, 57, 37–41, 2015.
47. Zhou, K., Zhang, Q.G., Han, G.L., Zhu, A.M., Liu, Q.L., Pervaporation of water–ethanol and methanol–MTBE mixtures using poly(vinyl alcohol)/cellulose acetate blended membranes. *J. Membr. Sci.*, 448, 93–101, 2013.
48. Bernal-ballén, A., Kuritka, I., Saha, P., Preparation and characterization of a bioartificial polymeric material: Bilayer of cellulose acetate-PVA. *J. Polym. Sci. A*, 2016, 1–11, 2016.
49. Nouri, M. and Marjani, A., Preparation of cellulose acetate membrane coated by PVA/Fe_3O_4 nanocomposite thin film: An *in situ* procedure. *Colloid Polym. Sci.*, 296, 1213–1223, 2018.
50. Petersen, N. and Gatenholm, P., Bacterial cellulose-based materials and medical devices: Current state and perspectives. *Appl. Microbiol. Biotechnol.*, 91, 1277–1286, 2011.
51. Yano, H., Optically transparent composites reinforced with networks of bacterial nanofibers. *Adv. Mater.*, 17, 153–155, 2005.
52. Hu, Y. and Catchmark, J.M., Formation and characterization of spherelike bacterial cellulose particles produced by *Acetobacter xylinum* JCM 9730 Strain. *Biomacromolecules*, 11, 1727–1734, 2010.
53. Jeon, S., Yoo, Y., Park, J., Kim, H., Hyun, J., Electrical conductivity and optical transparency of bacterial cellulose based composite by static and agitated methods. *Curr. Appl. Phys.*, 14, 1621–1624, 2014.
54. Barud, H.G.O., Barud, S., Cavicchioli, M., Silva, T., Batista, O., Junior, D.O., Santos, D.M., Luis, A., Almeida, D.O., Celes, F., Matos, V., De Oliveira, C.I., Francielli, P., Oliveira, D., Andrade, R., Crispim, D., Ribeiro, S.J.L., Preparation and characterization of a bacterial cellulose/silk fibroin sponge scaffold for tissue regeneration. *Carbohydr. Polym.*, 128, 41–51, 2015.
55. Panaitescu, D.M., Frone, A.N., Chiulan, I., Nanostructured biocomposites from aliphatic polyesters and bacterial cellulose. *Ind. Crops Prod.*, 93, 251–266, 2016.

56. Lu, M., Guan, X.-H., Xu, X.-H., Wei, D.-Z., Characteristic and mechanism of Cr(VI) adsorption by ammonium sulfamate-bacterial cellulose in aqueous solutions. *Chin. Chem. Lett.*, 24, 253–256, 2013.
57. Huang, Y., Zhu, C., Yang, J., Recent advances in bacterial cellulose. *Cellulose*, 21, 1–30, 2014.
58. Krystynowicz, A., Czaja, W., Wiktorowska, A., Gonc, M., Factors affecting the yield and properties of bacterial cellulose. *J. Ind. Microbiol. Biotechnol.*, 25, 189–195, 2002.
59. Guo, M., Ye, J., Gao, D., Xu, N., Yang, J., Agrobacterium-mediated horizontal gene transfer: Mechanism, biotechnological application, potential risk and forestalling strategy. *Biotechnol. Adv.*, 37, 259–270, 2019.
60. Li, J., Zhang, D., Que, Q., Chen, X., Ouyang, K., Industrial crops & products plant regeneration and Agrobacterium-mediated transformation of the miracle tree Neolamarckia cadamba. *Ind. Crops Prod.*, 130, 443–449, 2019.
61. Sarkar, R., A.K.C., Kumbhakar, P., Mandal, T., Aqueous synthesis and antibacterial activity of silver nanoparticles against pseudomonas putida. *Mater. Today Proc.*, 11, 686–694, 2019.
62. Granja-travez, R.S. and Bugg, T.D.H., Characterization of multicopper oxidase CopA from Pseudomonas putida KT2440 and Pseudomonas fluorescens Pf-5: Involvement in bacterial lignin oxidation. *Arch. Biochem. Biophys.*, 660, 97–107, 2018.
63. Sun, D., Zhuo, T., Hu, X., Fan, X., Zou, H., Identification of a Pseudomonas putida as biocontrol agent for tomato bacterial wilt disease. *Biol. Control*, 114, 45–50, 2017.
64. Tong, W., Li, X., Huo, Y., Zhang, L., Cao, Y., Wang, E., Chen, W., Tao, S., Wei, G., Genomic insight into the taxonomy of Rhizobium genospecies that nodulate Phaseolus vulgaris. *Syst. Appl. Microbiol.*, 41, 300–310, 2018.
65. Yotsombat, A., Hasegawa, T., Mino, K., Takata, G., Cloning and expression of D-glucoside 3-dehydrogenase from Rhizobium sp . S10 in Escherichia coli and its application for D -gulose production. *Protein Expr. Purif.*, 156, 58–65, 2019.
66. Flores-félix, J.D., Velázquez, E., García-fraile, P., Rhizobium and Phyllobacterium bacterial inoculants increase bioactive compounds and quality of strawberries cultivated in field conditions. *Food Res. Int.*, 111, 416–422, 2018.
67. Dimaio, M.A., Park, W.G., Longacre, T.A., Gastric Sarcina organisms in a patient with cystic fibrosis. *EHPC*, 1, 45–48, 2014.
68. Reiniati, I., Hrymak, A.N., Margaritis, A., Kinetics of cell growth and crystalline nanocellulose production by Komagataeibacter xylinus. *Biochem. Eng. J.*, 127, 21–31, 2017.

69. Guti, A. and Gonzalez-garcia, Y., XRD and solid state 13 C-NMR evaluation of the crystallinity enhancement of 13 C-labeled bacterial cellulose biosynthesized by Komagataeibacter xylinus under different stimuli: A comparative strategy of analyses. *Carbohydr. Res.*, 461, 51–59, 2018.
70. Chen, S., Lopez-sanchez, P., Wang, D., Mikkelsen, D., Gidley, M.J., Mechanical properties of bacterial cellulose synthesised by diverse strains of the genus Komagataeibacter. *Food Hydrocolloids*, 81, 87–95, 2018.
71. Brown, J., XLIIL-On an acetic ferment which form cellulose. *J. Chem. Soc., Trans.*, 49, 432–439, 1886.
72. Gomes, H., Barud, D.O., Rosa, R., Barud, S., Tercjak, A., Gutierrez, J., Rogério, W., Batista, O., Junior, D.O., Ribeiro, S.J.L., A multipurpose natural and renewable polymer in medical applications: Bacterial cellulose. *Carbohydr. Polym.*, 153, 406–420, 2016.
73. Qiao, K., Zheng, Y., Guo, S., Tan, J., Chen, X., Li, J., Xu, D., Wang, J., Hydrophilic nano fiber of bacterial cellulose guided the changes in the micro-structure and mechanical properties of nf-BC/PVA composites hydrogels. *Compos. Sci. Technol.*, 118, 47–54, 2015.
74. Wang, J., Gao, C., Zhang, Y., Wan, Y., Preparation and *in vitro* characterization of BC/PVA hydrogel composite for its potential use as artificial cornea biomaterial. *Mater. Sci. Eng. C*, 30, 214–218, 2010.
75. Abral, H., Hartono, A., Ha, F., Handayani, D., Sugiarti, E., Characterization of PVA/cassava starch biocomposites fabricated with and without sonication using bacterial cellulose fiber loadings. *Carbohydr. Polym.*, 206, 593–601, 2019.
76. Wong, S., Kasapis, S., Huang, D., Molecular weight and crystallinity alteration of cellulose via prolonged ultrasound fragmentation. *Food Hydrocolloids*, 26, 365–369, 2012.
77. Frone, A.N., Panaitescu, D.M., Donescu, D., Spataru, C.I., Radovici, C., Trusca, R., Somoghi, R., Preparation and characterization of PVA composites with cellulose nanofibers obtained by ultrasonication. *Bioresources*, 6, 487–512, 2011.
78. Id, M.A., Abral, H., Kasim, A., Pratoto, A., Mahardika, M., Mechanical properties of a water hyacinth nanofiber cellulose reinforced thermoplastic starch bionanocomposite: Effect of ultrasonic vibration during processing. *Fibers*, 6, 1–9, 2018.
79. Gea, S., Bilotti, E., Reynolds, C.T., Soykeabkeaw, N., Peijs, T., Bacterial cellulose – poly(vinyl alcohol) nanocomposites prepared by an *in-situ* process. *Mater. Lett.*, 64, 901–904, 2010.
80. Ciecha, D., Multifunctional bacterial cellulose/chitosan composite materials for medical applications. *Institute Chem. Fibres*, 12, 69–72, 2004.

81. Jipa, I.M., Dobre, L., Stroescu, M., Stoica-guzun, A., Jinga, S., Dobre, T., Preparation and characterization of bacterial cellulose-poly(vinyl alcohol) films with antimicrobial properties. *Mater. Lett.*, 66, 125–127, 2012.
82. Yudianti, R. and Karina, M., Development of nanocomposites from bacterial cellulose and poly(vinyl alcohol) using casting-drying method. *Proc. Chem.*, 4, 73–79, 2012.
83. Pavaloiu, R., Stoica-guzun, A., Stroescu, M., Jinga, S.I., Dobre, T., Composite films of poly(vinyl alcohol)- chitosan - bacterial cellulose for drug controlled release. *Int. J. Biol. Macromol.*, 68, 117–124, 2014.
84. Ramesh Babu, P. and Gaikar, V.G., Membrane characteristics as determinant in fouling of UF membranes. *Sep. Purif. Technol.*, 24, 23–34, 2001.
85. Pan, K., Zhang, X., Zhu, J., Cao, B., Grafting of regenerated cellulose membrane by surface-initiated atom transfer radical polymerization and its pH-resposive behavior. *Polym. Adv. Technol.*, 22, 1948–1952, 2011.
86. Biganska, O. and Navard, P., Morphology of cellulose objects regenerated from cellulose-N-methylmorpholine N-oxide-water solutions. *Cellulose*, 16, 179–188, 2009.
87. Ke, D., Liu, S., Dai, K., Zhou, J., Zhang, L., Peng, T., CdS/regenerated cellulose nanocomposite films for highly efficient photocatalytic H2 production under visible light irradiation. *J. Phys. Chem. C*, 113, 16021–16026, 2009.
88. Cuevas, A., Campos, B.B., Romero, R., Algarra, M., Vázquez, M.I., Benavente, J., Eco-friendly modification of a regenerated cellulose based film by silicon, carbon and N-doped carbon quantum dots. *Carbohydr. Polym.*, 206, 238–244, 2019.
89. Mohamed, M.A., Salleh, W.N.W., Jaafar, J., Ismail, A.F., Abd. Mutalib, M., Jamil, S.M., Feasibility of recycled newspaper as cellulose source for regenerated cellulose membrane fabrication. *J. Appl. Polym. Sci.*, 132, 1–10, 2015.
90. Cai, J. and Zhang, L., Rapid dissolution of cellulose in LiOH/urea and NaOH/urea aqueous solutions. *Macromol. Biosci.*, 5, 539–548, 2005.
91. Qi, H., Chang, C., Zhang, L., Effects of temperature and molecular weight on dissolution of cellulose in NaOH/urea aqueous solution. *Cellulose*, 15, 779–787, 2008.
92. Luo, X. and Zhang, L., New solvents and functional materials prepared from cellulose solutions in alkali/urea aqueous system. *Food Res. Int.*, 52, 387–400, 2013.
93. Gao, Q., Shen, X., Lu, X., Regenerated bacterial cellulose fibers prepared by the NMMO·H2O process. *Carbohydr. Polym.*, 83, 1253–1256, 2011.
94. Mohamed, M.A., Salleh, W.N.W., Jaafar, J., Ismail, A.F., Abd Mutalib, M., Mohamad, A.B., Zain, M.F., Awang, N.A., Mohd Hir, Z.A., Physicochemical characterization of cellulose nanocrystal and nanoporous self-assembled

CNC membrane derived from Ceiba pentandra. *Carbohydr. Polym.*, 157, 1892–1902, 2017.
95. Qi, H., Cai, J., Zhang, L., Kuga, S., Properties of films composed of cellulose nanowhiskers and a cellulose matrix regenerated from alkali/urea solution. *Biomacromolecules*, 10, 1597–602, 2009.
96. Mohamed, M.A., Salleh, W.N.W., Jaafar, J., Ismail, A.F., Mutalib, M.A., Sani, N.A.A., Asri, S.E.A.M., Ong, C.S., Physicochemical characteristic of regenerated cellulose/N-doped TiO2 nanocomposite membrane fabricated from recycled newspaper with photocatalytic activity under UV and visible light irradiation. *Chem. Eng. J.*, 284, 202–215, 2016.
97. Mohamed, M.A., Salleh, W.N.W., Jaafar, J., Asri, S.E.A.M., Ismail, A.F., Physicochemical properties of "green" nanocrystalline cellulose isolated from recycled newspaper. *RSC Adv.*, 5, 29842–29849, 2015.
98. Cazón, P., Velazquez, G., Vázquez, M., Novel composite films from regenerated cellulose-glycerol-polyvinyl alcohol: Mechanical and barrier properties. *Food Hydrocolloids*, 89, 481–491, 2019.
99. Zhang, X., Zhu, J., Liu, X., Preparation and characterization of regenerated cellulose blend films containing high amount of poly(vinyl alcohol) (PVA) in ionic liquid. *Macromol. Res.*, 20, 703–708, 2012.
100. Lu, Y., Liu, Y., Yu, H., Sun, Q., Preparation, characterization and biocompatibility of regenerated cellulose/PVA blend membranes in 1-allyl-3-methyimidazolium chloride. *Adv. Mater. Res.*, 180, 1108–1111, 2011.
101. Abdulkhani, A., Hojati, E., Ashori, A., Hamzeh, Y., Naghi, A., Preparation of cellulose/polyvinyl alcohol biocomposite films using 1-n-butyl-3-methylimidazolium chloride. *Int. J. Biol. Macromol.*, 62, 379–386, 2013.
102. Kaco, H., Zakaria, S., Chia, C.H., Zhang, L., Transparent and printable regenerated kenaf cellulose/PVA film. *Bioresources*, 9, 2167–2178, 2014.
103. Dai, H., Zhang, H., Ma, L., Zhou, H., Yu, Y., Guo, T., Green pH/magnetic sensitive hydrogels based on pineapple peel cellulose and polyvinyl alcohol: Synthesis, characterization and naringin prolonged release. *Carbohydr. Polym.*, 209, 51–61, 2019.
104. K.L.T.C., Tamayo, G., Donohue, A., Kobayashi, T., R.A.S.A., Obtaining of hydrogels using PVA and HEC for adipose tissue regeneration. *J. Tissue Sci. Eng.*, 6, 1–5, 2015.
105. Jayaramudu, T., Ko, H., Kim, H.C., Kim, J.W., Muthoka, R.M., Kim, J., Electroactive hydrogels made with polyvinyl alcohol/cellulose nanocrystals. *Materials*, 11, 1–11, 2018.
106. Kamoun, E.A., Chen, X., Mohy, M.S., Kenawy, E.S., Crosslinked poly(vinyl alcohol) hydrogels for wound dressing applications: A review of remarkably blended polymers. *Arab. J. Chem.*, 8, 1–14, 2015.

107. Rui-hong, X., Peng-gang, R., Jian, H., Fang, R., Lian-zhen, R., Preparation and properties of graphene oxide-regenerated cellulose/polyvinyl alcohol hydrogel with pH-sensitive behavior. *Carbohydr. Polym.*, 138, 222–228, 2016.
108. Chang, C., Lue, A., Zhang, L., Effects of crosslinking methods on structure and properties of cellulose/PVA hydrogels. *Macromol. Chem. Phys.*, 209, 1266–1273, 2008.
109. Ioana, R., Bercea, M., Avadanei, M., Lisa, G., Biliuta, G., Coseri, S., Green route for the fabrication of self-healable hydrogels based on tricarboxy cellulose and poly(vinyl alcohol). *Int. J. Biol. Macromol.*, 123, 744–751, 2019.
110. Bai, H., Li, Z., Zhang, S., Wang, W., Dong, W., Interpenetrating polymer networks in polyvinyl alcohol/cellulose nanocrystals hydrogels to develop absorbent materials. *Carbohydr. Polym.*, 200, 468–476, 2018.
111. Dong, S., Bortner, M.J., Roman, M., Analysis of the sulfuric acid hydrolysis of wood pulp for cellulose nanocrystal production: A central composite design study. *Ind. Crops Prod.*, 93, 76–87, 2016.
112. Rowell, B.Y.R.M., Acetylation of wood. *For. Prod. J.*, 56, 4–12, 2006.
113. Paulo, J., Morais, S., De Freitas, M., De, M., De Souza, M., Dias, L., Magalhães, D., Ribeiro, A., Extraction and characterization of nanocellulose structures from raw cotton linter. *Carbohydr. Polym.*, 91, 229–235, 2013.
114. Cao, J., Sun, X., Lu, C., Zhou, Z., Zhang, X., Yuan, G., Water-soluble cellulose acetate from waste cotton fabrics and the aqueous processing of all-cellulose composites. *Carbohydr. Polym.*, 149, 60–67, 2016.
115. Mohamed, M.A., Salleh, W.N.W., Jaafar, J., Ismail, A.F., Abd Mutalib, M., Mohamad, A.B., Zain, M.F.M., Awang, N.A., Mohd Hir, Z.A., Physicochemical characterization of cellulose nanocrystal and nanoporous self-assembled CNC membrane derived from Ceiba pentandra. *Carbohydr. Polym.*, 157, 1892–1902, 2016.
116. Xiong, R., Zhang, X., Tian, D., Zhou, Z., Lu, C., Comparing microcrystalline with spherical nanocrystalline cellulose from waste cotton fabrics. *Cellulose*, 19, 1189–1198, 2012.
117. Jahan, Z., Bilal, M., Niazi, K., Weiby, Ø., Mechanical , thermal and swelling properties of cellulose nanocrystals/PVA nanocomposites membranes. *J. Ind. Eng. Chem.*, 57, 113–124, 2018.
118. Rathore, A. and Pradhan, M.K., Hybrid cellulose bionanocomposites from banana and jute fibre: A review of preparation, properties and applications. *Mater. Today Proc.*, 4, 3942–3951, 2017.
119. Deng, L. and Hägg, M., Swelling behavior and gas permeation performance of PVAm/PVA blend FSC membrane. *J. Membr. Sci.*, 363, 295–301, 2010.
120. Cha, R., Wang, C., Cheng, S., He, Z., Jiang, X., Using carboxylated nanocrystalline cellulose as an additive in cellulosic paper and poly(vinyl alcohol) fiber paper. *Carbohydr. Polym.*, 110, 298–301, 2014.

121. Babu, A., Selvam, P., Nambiar, R.B., Rotimi, E., Development of polyvinyl alcohol/chitosan bio-nanocomposite films reinforced with cellulose nanocrystals isolated from rice straw. *Appl. Surf. Sci.*, 449, 591–602, 2018.
122. Noshirvani, N., Hong, W., Ghanbarzadeh, B., Fasihi, H., Study of cellulose nanocrystal doped starch-polyvinyl alcohol bionanocomposite films. *Int. J. Biol. Macromol.*, 107, 2065–2074, 2018.
123. Karim, Z., Claudpierre, S., Grahn, M., Oksman, K., Mathew, A.P., Nanocellulose based functional membranes for water cleaning: Tailoring of mechanical properties, porosity and metal ion capture. *J. Membr. Sci.*, 514, 418–428, 2016.
124. Yang, R., Aubrecht, K.B., Ma, H., Wang, R., Grubbs, R.B., Hsiao, B.S., Chu, B., Thiol-modified cellulose nanofibrous composite membranes for chromium (VI) and lead (II) adsorption. *Polym. (Guildf)*, 55, 1167–1176, 2014.
125. Lee, S., Mohan, D.J., Kang, I., Doh, G., Lee, S., Han, S.O., Nanocellulose reinforced PVA composite films: Effects of acid treatment and filler loading. *Fibers Polym.*, 10, 77–82, 2009.
126. Kakroodi, A.R., Cheng, S., Sain, M., Asiri, A., Mechanical, thermal, and morphological properties of nanocomposites based on polyvinyl alcohol and cellulose nanofiber from Aloe vera rind. *J. Nanomater.*, 2014, 12–18, 2014.
127. Qua, E.H., Hornsby, P.R., Sharma, H.S.S., Lyons, G., Mccall, R.D., Preparation and characterization of poly(vinyl alcohol) nanocomposites made from cellulose nanofibers. *J. Appl. Polym. Sci.*, 113, 2238–2247, 2009.
128. Ben Cheikh, S., Ben Cheikh, R., Cunha, E., Lopes, P.E., Ben, S., Ben Cheikh, R., Cunha, E., Paulo, E., Production of cellulose nanofibers from Alfa grass and application as reinforcement for polyvinyl alcohol. *Plast. Rubber Compos.*, 47, 297–305, 2018.
129. Poyraz, B., Tozluo, A., Candan, Z., Demir, A., Yavuz, M., Influence of PVA and silica on chemical, thermo-mechanical and electrical properties of Celluclast-treated nanofibrillated cellulose composites. *Int. J. Biol. Macromol.*, 104, 384–392, 2017.
130. Bian, H., Wei, L., Lin, C., Dai, H., Zhu, J.Y., Lignin-containing cellulose nanofibril-reinforced polyvinyl alcohol hydrogels. *ACS Sustainable Chem.*, 4, 4821–4828, 2018.
131. Irvin, C.W., Satam, C.C., Meredith, J.C., Shofner, M.L., Mechanical reinforcement and thermal properties of PVA tricomponent nanocomposites with chitin nanofibers and cellulose nanocrystals. *Compos. Part A*, 116, 147–157, 2019.
132. Puttaswamy, M. and Srinikethan, G., Biocomposite composed of PVA reinforced with cellulose microfiber isolated from biofuel industrial dissipate: Jatropha Curcus L. seed shell. *J. Environ. Chem. Eng.*, 5, 1990–1997, 2017.

133. Zhang, J., Lei, W., Chen, J., Liu, D., Tang, B., Li, J., Wang, X., Enhancing the thermal and mechanical properties of polyvinyl alcohol (PVA) with boron nitride nanosheets and cellulose nanocrystals. *Polym. (Guildf)*, 148, 101–108, 2018.
134. Chowdhury, R.A., Rai, A., Glynn, E., Morgan, P., Moore, A.L., Youngblood, P., Superior, processing-dependent thermal conductivity of cellulose nanocrystal-poly(vinyl alcohol) composite films. *Polymer*, 164, 17–25, 2019.
135. Mandal, A. and Chakrabarty, D., Studies on the mechanical, thermal, morphological and barrier properties of nanocomposites based on poly(vinyl alcohol) and nanocellulose from sugarcane bagasse. *J. Ind. Eng. Chem.*, 20, 462–473, 2014.
136. Bunaciu, A.A., Udriştioiu, E., Aboul-enein, H.Y., Bunaciu, A.A., Udriştioiu, E., Aboul-enein, H.Y., Bunaciu, A.A., E.G.U.S., X-Ray diffraction: Instrumentation and applications. *Crit. Rev. Anal. Chem.*, 45, 289–299, 2015.
137. Jacob, J., Peter, G., Thomas, S., Haponiuk, J.T., Gopi, S., Chitosan and polyvinyl alcohol nanocomposites with cellulose nano fibers from ginger rhizomes and its antimicrobial activities. *Int. J. Biol. Macromol.*, 129, 370–376, 2019.
138. Korhonen, O., Zrida, M., Hamzaoui, A.H., Budtova, T., All-cellulose composites from alfa and wood fibers. *Ind. Crops Prod.*, 127, 135–141, 2019.
139. Wang, Z., Qiao, X., Sun, K., Rice straw cellulose nanofibrils reinforced poly (vinyl alcohol) composite films. *Carbohydr. Polym.*, 197, 442–450, 2018.
140. Asgher, M., Ahmad, Z., Iqbal, H.M.N., Bacterial cellulose-assisted delignified wheat straw-PVA based bio-composites with novel characteristics. *Carbohydr. Polym.*, 161, 244–252, 2017.
141. Jahan, Z., Bilal, M., Niazi, K., Hägg, M., Weiby, Ø., Decoupling the effect of membrane thickness and CNC concentration in PVA based nanocomposite membranes for CO_2/CH_4 separation. *Sep. Purif. Technol.*, 204, 220–225, 2018.
142. Ahn, Y., Song, Y., Kwak, S.-Y., Kim, H., Highly ordered cellulose II crystalline regenerated from cellulose hydrolyzed by 1-butyl-3-methylimidazolium chloride. *Carbohydr. Polym.*, 137, 321–327, 2016.
143. Liu, Z., Wang, H., Li, Z., Lu, X., Zhang, X., Zhang, S., Zhou, K., Characterization of the regenerated cellulose films in ionic liquids and rheological properties of the solutions. *Mater. Chem. Phys.*, 128, 220–227, 2011.
144. Chen, X., Zhang, Y., Wang, H., Wang, S., Liang, S., Ralph, H., Chen, X., Zhang, Y., Wang, H., Solution rheology of cellulose in 1-butyl-3-methyl imidazolium chloride. *J. Rheol.*, 55, 485–494, 2011.
145. Zhou, L., He, H., Li, M., Song, K., Cheng, H.N., Wu, Q., Morphological influence of cellulose nanoparticles (CNS) from cottonseed hulls on rheological properties of polyvinyl alcohol/CN suspensions. *Carbohydr. Polym.*, 153, 445–454, 2016.

146. Singh, S., Gaikwad, K.K., Lee, Y.S., Antimicrobial and antioxidant properties of polyvinyl alcohol bio composite films containing seaweed extracted cellulose nano-crystal and basil leaves extract. *Int. J. Biol. Macromol.*, 107, 1879–1887, 2018.
147. Moawad, S.A., El-ghorab, A.H., Hassan, M., Nour-eldin, H., Chemical and microbiological characterization of egyptian cultivars for some spices and herbs commonly exported abroad. *Food Sci. Nutr.*, 6, 643–659, 2015.
148. Ngamakeue, N. and Chitprasert, P., Encapsulation of holy basil essential oil in gelatin: Effects of palmitic acid in carboxymethyl cellulose emulsion coating on antioxidant and antimicrobial activities. *Food Bioprocess Technol.*, 9, 1735–1745, 2016.
149. Vaccinium, L. and Kra, K., Effect of pullulan coating on postharvest quality and shelf-life of highbush blueberry. *Materials*, 10, 1–17, 2017.
150. Cerpakovska, D., Kalnins, M., Tupureina, V., Biodegradable paper reinforced poly (vinyl alcohol) composites for packaging. *Sci. J. Riga Tech. Univ.*, 21, 26, 2010.
151. Cinelli, P., Chiellini, E., Lawton, J.W., Imam, S.H., Foamed articles based on potato starch, corn fibers and poly(vinyl alcohol). *Polym. Degrad. Stab.*, 91, 1147–1155, 2006.
152. Awang, N.A., Salleh, W.N.W., Ismail, A.F., Yusof, N., Aziz, F., Jaafar, J., Adsorption behavior of chromium(vi) onto regenerated cellulose membrane. *Ind. Eng. Chem. Res.*, 58, 720–728, 2019.
153. Hadi, M., Sarmadi, M., Raza, M., Investigating the equilibrium and adsorption kinetics for the removal of Ni (II) ions from aqueous solutions using adsorbents prepared from the modified waste newspapers: A low-cost and available adsorbent. *Microchem. J.*, 146, 1043–1053, 2019.
154. Farag, A.M., Sokker, H.H., Zayed, E.M., Nour, F.A., Abd, N.M., Removal of hazardous pollutants using bifunctional hydrogel obtained from modified starch by grafting copolymerization. *Int. J. Biol. Macromol.*, 120, 2188–2199, 2018.
155. Ahmad, A., Mohd-Setapar, S.H., Chuong, C.S., Khatoon, A., Wani, W.A., Kumar, R., Rafatullah, M., Recent advances in new generation dye removal technologies: Novel search for approaches to reprocess wastewater. *RSC Adv.*, 5, 30801–30818, 2015.
156. Jain, D.S.M.C.K. and Yadav, A.K., Removal of heavy metals from emerging cellulosic low-cost adsorbents: A review. *Appl. Water Sci.*, 7, 2113–2136, 2017.
157. Saha, B. and Orvig, C., Biosorbents for hexavalent chromium elimination from industrial and municipal effluents. *Coord. Chem. Rev.*, 254, 2959–2972, 2010.

158. Wong, K.C., Goh, P.S., Ismail, A.F., Gas separation performance of thin film nanocomposite membranes incorporated with polymethyl methacrylate grafted multi-walled carbon nanotubes. *Int. Biodeterior. Biodegradation*, 102, 339–345, 2015.
159. Muhamad, N., Abdullah, N., Rahman, M.A., Abas, K.H., Aziz, A.A., Removal of nickel from aqueous solution using supported zeolite-Y hollow fiber membranes. *Environ. Sci. Pollut. Res.*, 25, 19054–19064, 2018.
160. Hosseini, S.S., Bringas, E., Tan, N.R., Ortiz, I., Ghahramani, M., Alaei Shahmirzadi, M.A., Recent progress in development of high performance polymeric membranes and materials for metal plating wastewater treatment: A review. *J. Water Process Eng.*, 9, 78–110, 2016.
161. Du, X., Qu, F., Liang, H., Li, K., Yu, H., Bai, L., Li, G., Removal of antimony (III) from polluted surface water using a hybrid coagulation–flocculation–ultrafiltration (CF–UF) process. *Chem. Eng. J.*, 254, 293–301, 2014.
162. Fu, F. and Wang, Q., Removal of heavy metal ions from wastewaters: A review. *J. Environ. Manage.*, 92, 407–418, 2011.
163. Aziz, H.A., Adlan, M.N., Ariffin, K.S., Heavy metals (Cd, Pb, Zn, Ni, Cu and Cr (III)) removal from water in Malaysia: Post treatment by high quality limestone. *Bioresour. Technol.*, 99, 1578–1583, 2008.
164. Zhao, M., Xu, Y., Zhang, C., Rong, H., Zeng, G., New trends in removing heavy metals from wastewater. *Appl. Microbiol. Biotechnol.*, 100, 6509–6518, 2016.
165. Gunatilake, S.K., Methods of removing heavy metals from industrial wastewater. *J. Multidiscip. Eng. Sci. Stud.*, 1, 12–18, 2015.
166. Larous, S., Meniai, A., Lehocine, M.B., Experimental study of the removal of copper from aqueous solutions by adsorption using sawdust. *Desalination*, 185, 483–490, 2005.
167. Chen, Z., Deng, M., Chen, Y., He, G., Wu, M., Wang, J., Preparation and performance of cellulose acetate/polyethyleneimine blend microfiltration membranes and their applications. *J. Membr. Sci.*, 235, 73–86, 2004.
168. Garg, U.K., Kaur, M.P., Garg, V.K., Sud, D., Removal of hexavalent chromium from aqueous solution by agricultural waste biomass. *J. Hazard. Mater.*, 140, 60–68, 2007.
169. Pehlivan, E., Biosorption of chromium (VI) ion from aqueous solutions using walnut , hazelnut and almond shell. *J. Hazard. Mater.*, 155, 378–384, 2008.
170. Lei, C., Gao, J., Ren, W., Xie, Y., Abdalkarim, S.Y.H., Wang, S., Ni, Q., Yao, J., Fabrication of metal-organic frameworks cellulose aerogels composite materials for removal of heavy metal ions in water. *Carbohydr. Polym.*, 205, 35–41, 2019.

171. Tran, T.H., Okabe, H., Hidaka, Y., Hara, K., Removal of metal ions from aqueous solutions using carboxymethyl cellulose/sodium styrene sulfonate gels prepared by radiation grafting. *Carbohydr. Polym.*, 157, 335–343, 2017.
172. Wan Ngah, W.S., Teong, L.C., Hanafiah, M.A.K.M., Adsorption of dyes and heavy metal ions by chitosan composites: A review. *Carbohydr. Polym.*, 83, 1446–1456, 2011.
173. Liu, P., Oksman, K., Mathew, A.P., Surface adsorption and self-assembly of Cu(II) ions on TEMPO-oxidized cellulose nanofibers in aqueous media. *J. Colloid Interface Sci.*, 464, 175–182, 2016.
174. Zhu, Q., Wang, Y., Li, M., Liu, K., Hu, C., Yan, K., Sun, G., Wang, D., Activable carboxylic acid functionalized crystalline nanocellulose/PVA-co-PE composite nanofibrous membrane with enhanced adsorption for heavy metal ions. *Sep. Purif. Technol.*, 186, 70–77, 2017.
175. Wang, L. and Wang, M., Removal of heavy metal ions by poly(vinyl alcohol) and carboxymethyl cellulose composite hydrogels prepared by a freeze – thaw method. *ACS Sustainable Chem. Eng.*, 4, 2830–2837, 2016.
176. She, J., Tian, C., Wu, Y., Li, X., Cellulose nanofibrils aerogel cross-linked by poly(vinyl alcohol) and acrylic acid for efficient and recycled adsorption with heavy metal ions. *J. Nanosci. Nanotechnol.*, 18, 4167–4175, 2018.
177. Adsorption, I., Wang, D., Cheng, W., Yue, Y., Xuan, L., Ni, X., Han, G., Electrospun cellulose nanocrystals/chitosan/polyvinyl alcohol nanofibrous films and their exploration to metal. *Polymers*, 10, 1–17, 2018.
178. Zheng, Q. and Gong, S., Green synthesis of polyvinyl alcohol (PVA)-cellulose nanofibril (CNF) hybrid aerogels and their use as superabsorbents. *J. Mater. Chem. A*, 2, 3110–3118, 2014.

5

Polyvinyl Alcohol/Starch-Based Biocomposites and Bionanocomposites

Nor Fasihah Binti Zaaba* **and Hanafi Bin Ismail**

School of Materials and Mineral Resources Engineering, Universiti Sains Malaysia Nibong Tebal, Penang, Malaysia

Abstract

In general, petrochemical-based plastics (polyolefins, polyesters, and polyamides) have been used as packaging materials due to its superior properties such as durable and low in cost. Nevertheless, plastic is non-degradable material and the extensive use of plastics lead to rise of the waste management problems and environmental pollution. Thus, biodegradable plastic is an ideal alternative to replace the non-degradable plastic material. Among biodegradable plastics, polyvinyl alcohol (PVA) is chosen as the most significant plastic materials as it offers excellent mechanical properties. Normally, starch is the best options to blend with PVA to reduce the material's cost while maintaining the high properties and characteristics of PVA. In this article, the preparation, characterization, mechanical properties and application of PVA/starch biocomposites and bionanocomposites from previous studies are studied, briefly.

Keywords: Polyvinyl alcohol, biocomposites, nanocomposites, mechanical properties

5.1 Introduction

Owing to the distinctive properties and characteristics such as widely available at low cost, lightweight, durable, good barrier properties to oxygen and aroma compounds, good tensile and tear strength, and heat sealability,

**Corresponding author*: fasehazaaba@gmail.com

Visakh P. M. and Olga B. Nazarenko (eds.) Polyvinyl Alcohol-Based Biocomposites and Bionanocomposites, (131–150) © 2023 Scrivener Publishing LLC

the fabrication of petrochemical based plastics (polyolefins, polyesters, and polyamides) have been broadly used as packaging materials [1, 2]. However, the accumulation of plastic in the environment is one of the main problems that we are facing today. The consumption of plastic leads to rise of the waste management problems and pollution. Thus, one of the preferences to deal with such problems is the degradation of waste plastics by several resources [3]. The degradation of these materials starts with the radiation of solar ultraviolet (UV) that absorbs by a variety of synthetic polymers. Then, these materials go through photolytic, photo-oxidative, and thermo-oxidative reactions [4].

The addition of some additives increased the tendency of polymeric materials to experience solar UV radiation and eventually led to degradation [5–8]. However, the degradation of these plastic is tedious and time-consuming [7, 9, 10]. Indeed, these plastics are biodegradation-resistant when released into the environment and tends to accumulate from year to year [11–15]. The anxieties of these environmental issues have motivated the determinations on fabrication and improvement of green materials from recyclable and biodegradable materials [8, 16–20]. Thus, there have been a lot of intensive researches regarding the degradability and the development of traditional plastics to novel biodegradable polymers. Criteria for harmless and cleaner environment have focused to eco-composite materials that can be bio-assimilated or degraded [21]. Biocomposites or bionanocomposites that prepared totally from renewable resources are immediately essential to develop the degradability and material recyclability [22].

5.2 Polyvinyl Alcohol/Starch-Based Biocomposites and Bionanocomposites

These days, polyvinyl alcohol (PVA) is the most significant plastic materials used in the industrial among biodegradable polymers [23]. It is an ecologically biodegradable polymer with excellent mechanical properties [24]. Several studies have been reported on water soluble PVA, where it has been found to be mineralized and biodegraded in numerous environments condition [25]. The application of PVA is mostly implemented in agricultural mulch films, paper coatings and packaging application. Besides, PVA is also used as lubricant to make biodegradable carriers or water soluble in eye drops and hard contact lens solution [26]. The biodegradability of PVA in various microbial environments has been widely reported [27–32].

Starch is one of the best options to be blended with PVA [33, 34]. Starch is made up of two types of materials, amylose, a linear chain molecule of α-1,4-linked D-glucose and amylopectin, a branched polymer of α-1,4-linked D-glucose with a 1,6-linked D-glucose unit. It is reflected as a condensation polymer of glucose. In fact, it is capable to be biodegraded by a number of enzymes through enzyme catalyzed hydrolysis [35]. The PVA/starch blends have shown an excellent compatibility [36, 37]. In fact, PVA and starch exhibited highly desirable physical blend in various applications. Blend of these PVA-based materials with starches permits the manufacturing of water-soluble blends at lower cost while maintaining the high properties and characteristics of PVA. Several studies have reported their patented formulation of PVA/starch blend showed an enhancement in strength, modulus as well as compatibility of the blend [24, 33]. However, their tensile strength, elongation at break as well as their water resistance properties need to further increase [27, 30, 38, 39]. The compatibility between starch and synthetic polymers can be improved by the addition of compatibilizer or chemical modification [40–43]. Apart from that, cross-linking, etherification, esterification and oxidation also can be applied for starch modification. Above all, methylation was usually chosen as one of the esterification methods in interpreting the substitution pattern in polymer chains and the structure of polysaccharides. They have been verified as an effective technique to develop the blend's properties [37, 44]. In another way, crosslinking agent also can be added to improve the physical and mechanical properties of the starch/PVA blends [45].

In term of biodegradability, the PVA/starch blends have lower biodegradability than other biodegradable plastics. As reported by Bastioli *et al.* [46] on amylose-PVA composites (PVA-starch blend) as about 75% weight loss was detected in 300 days of degradation test with activated sludge. Besides that, Iwanami *et al.* [47] also stated their findings on PVA-starch blend, only 50% weight loss was detected in a waste-composting process. In another finding, Tudorachi *et al.* [48] also reported a significant weight loss was detected as well as a decrease of the physical–mechanical characteristics based in the presence of microorganisms for combination of PVA, starch, glycerine and urea polymer materials. The factors that affect the degradation of the polymeric materials are the chemical composition of the blend, molecular weight, hydrophobic character, size of the molecules, additives used and the environmental conditions. Indeed, the higher the content of starch contributes the highest values of weight loss of the samples [12, 49].

5.3 Preparation

According to Zaaba *et al.* [50], the PVA/tapioca starch blend was prepared using the solution casting method. Firstly, the PVA were poured into a round bottom flask containing de-ionized water. Then, the solution was stirred with high speed (>1000 rpm) for 10 min in a water bath at temperature 95°C. Next, the tapioca starch was dissolved stirred for another 10 min. Lastly, the additives were added and stirring was continued for another 10 min (>1000 rpm). The mixtures were cast onto a glass plate which was placed on a flat surface. The films were dried at room temperature for 12 h before cured in oven at 90°C for 30 min. Then, the films were removed from the glass plate and kept for mechanical testing.

While for the preparation of PVA/starch blend via solution casting by Denis *et al.* [51], the preparation was begun with PVA, starch and additives were dispersed in distilled water. Then, by using a mechanical stirrer (VELP Scientifica, Italy), the blend was stirred at 2000 rpm for 15 min. To complete the cross-linking reaction, the mixture needs to be heated up to 90 °C under stirring till gelatinization for another 45 min at 98 °C Then, it was poured on PET plates and drying at ambient temperature overnight.

Besides that, PVA/starch blend was also prepared by Susmita Dey Sadhu *et al.* [52] where the PVA/starch mixture was firstly heated at 70°C until uniformity appears. Then, the solutions were cast on a casting mold and dried in an oven at 75°C. Another blend preparation was reported by Zhijun Wu *et al.* [53] on PVA/starch blend. Firstly, the PVA, starch and glycerol were dissolved in distilled water and the blend was stirred using an electric stirrer for 45 min at 95°C. Next, the solutions were poured onto the glass plates and dried for 24 h at room temperature. The dried films were removed from the glass plates and kept for further testing.

Apart from that, Xiong *et al.* [54] also stated the preparation of PVA/corn starch blend in their study. The mixture of PVA, corn starch and de-ionised water were firstly poured into a round flask. Then, in a water bath, the mixture was stirred with a high speed mixer for 30 minutes at a constant temperature. After that, the additives were added into the mixture and stirred for 10 minutes till a smooth film surface was shown. Next, the mixture was poured onto a glass plate that was placed on a flat surface. The blend was dried in an enclosed environment for almost 12 hours and then cured in the oven for 30 minutes. Lastly, the films were peeled off from the glass plate.

5.4 Characterizations

The mechanical properties and characterization of PVA/starch blend are tested and measured based on the behaviour of the blend influenced by different mode of deterioration and the nature of material used.

5.4.1 Mechanical Properties

Mechanical properties define a material's behaviour once subjected to mechanical stresses. The properties include tensile strength, elongation at break and tensile modulus measurements. As reported by Zaaba *et al.* [50] the tensile strength (TS), elongation at break (Eb) and tensile modulus of PVA/tapioca starch biodegradable films were evaluated according to ASTM D882 using the Instron 3366 testing machine. Each film was cut with about 0.09 mm average thickness and 6.4 mm width. In another study, Shangwen Thang *et al.* [55] reported that the tensile strength and elongation at break of the PVA/starch/nano-silicon dioxide (nano-SiO_2) biodegradable blend films were measured according to Chinese standard method GB/T4456-96 (Polyethylene Blown film for packaging, 1996) using the electron tensile tester CMT-6104 (Shenzhen Sans Test Machine Co., Ltd., China). For the past few years, numerous studies have been performed on PVA/starch blends, in order to characterize its mechanical properties [34, 50–52, 55].

For example, Shangwen Thang *et al.* [5] reported the tensile strength and elongation at break of PVA/starch/nano-SiO_2 biodegradable blend films at different nano-SiO_2 content as shown in Figure 5.1. The tensile strength of PVA/starch without nano-SiO_2 was 9.03 MPa. As the content of nano-SiO_2 increase, the tensile strength of the blend also increased (15.0 MPa), up to touched at maximum point which was about 2.5 wt.% of the nano-SiO_2 content. After that point, the tensile strength of the blend film started to decrease along with the increment of nano-SiO_2 content. However, the value was still far greater than that of the blend without nano-SiO_2.

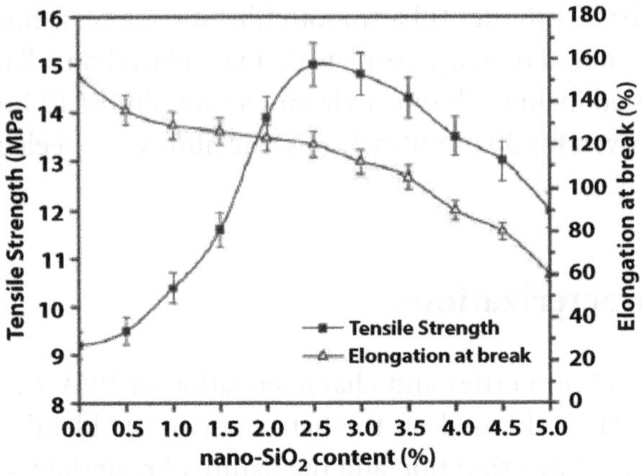

Figure 5.1 Tensile strength and elongation at break of starch/PVA/nano-SiO_2 biodegradable blend films [55].

The significant rise in the tensile strength of the blend films verified the existence of intermolecular interactions between the nano-SiO_2 and the starch or the nano-SiO_2 and the PVA in the blend films. As for elongation at break of PVA/starch/nano-SiO_2 blend, the value was decreased gradually along with an increase in nano-SiO_2 content.

In another example, Denis *et al.* [51] reported on PVA/starch blend films with different concentration of cellulose nanofibers (CN). Figure 5.2 shows the elongation at break, tensile strength, and modulus of elasticity of PVA/starch blend films with different amount of CN and storage conditioned. There were 3 sets of conditions applied before characterization by tensile tests and XRD: (I) the films were stored at room conditions for 15 days, (II) 30 days at room conditions followed by thermal treatment at 60 °C for 24 h and then in a desiccator at 25 °C for 24 h, (III) 30 days at room conditions and at 25 °C, RH 75% for another 15 days. Obviously, the elongation at break for all samples was small due to the cross-linking between the PVA and the starch. Thus, decreases the ductility of the films [56]. The incorporation of CN increases the films rigidity subsequently altered the elongation at break into a small amount, irrespective of storage conditions. Besides that, the tensile strength also affected by the concentration of CN as the increment detected for (I) and (II) storage conditions. However, for (III) storage conditions, the tensile strength only increased to 4 wt.% of CN concentration. For modulus of elasticity, similar trend of graph was observed. Additionally, both tensile strength and modulus of elasticity

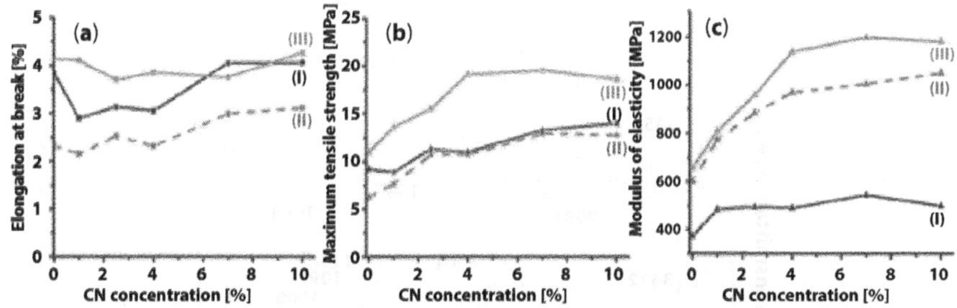

Figure 5.2 Mechanical properties of S/PVA/CN films as functions of CN concentration: (a) elongation at break; (b) maximum tensile strength; (c) modulus of elasticity.

displayed a slightly increment at more than 4 wt.% of CN, for all the storage conditions. This was might be due to the CN network formed in the blend at more than 4 wt.% of CN.

5.4.2 Fourier Transform Infrared (FTIR) Spectroscopy

FTIR spectroscopy is another important procedure for investigating the chemical changes of polymer. It is use for both qualitative and quantitative characterization of chemical changes and degradation materials [50, 51, 57, 58]. The formation of several functional groups from chemical changes are intensely reliant on the chemical structure of the polymer. The FTIR spectra is recorded on a Perkin-Elmer FTIR spectrometer. The wave range is from 4000 to 550 cm^{-1} and the scanning resolution is 4 cm^{-1}. For film sample preparation, the films were cut into small 10 mm X 10 mm pieces from each sample. Then, it was placed on Miracle ATR accessory (miracle base optics assembly) and the spectra is recorded in reflection.

Figure 5.3 shows the IR spectra of PVA, starch (ST) and nano-SiO_2, as reported by Shangwen Thang *et al.* [55]. Obviously, there was a wide and strong absorption band of the nano-SiO_2 at 3452 cm^{-1}. This showed that there were a plenty of –OH groups on the surface of nano-SiO_2. While the antisymmetric stretching vibration, symmetric stretching vibration, and flexural vibration of Si-O-Si were detected at 1088 cm^{-1}, 797 cm^{-1}, and 459 cm^{-1}, respectively. For the IR spectra of ST, a broad and strong absorption peak at 3412 cm^{-1} can be seen indicated the characteristic absorption peak of the stretching vibration of -OH and the hydrogen bonds' association in -OH groups. Further, the band at 1158 cm^{-1} and 1091 cm^{-1} were assigned to CO and COH groups while the band at 1000 cm^{-1} was attributed to

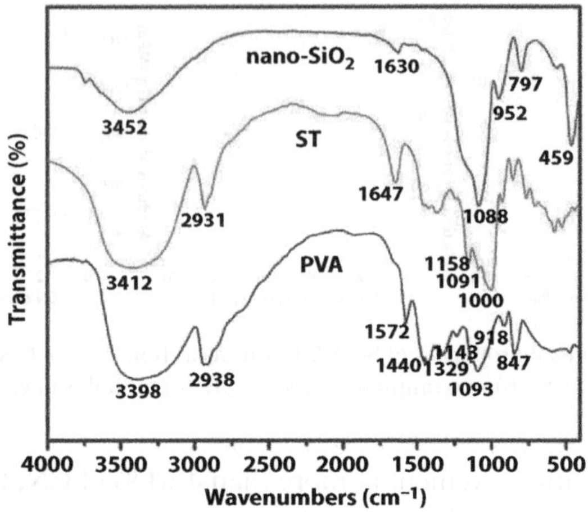

Figure 5.3 The IR spectra of nano-SiO_2, ST, and PVA [5].

the stretching vibration of CO in the COC groups. As compared with almost organic compound, an absorption band of PVA can be seen at 2938 cm^{-1} attributed to the stretching vibrations of the CH and CH2 groups. Meanwhile, the broad hydroxyl band existed at 3000–3600 cm^{-1}, and the CO stretching occurred at 1000–1260 cm^{-1}.

In another study, Judawisastra *et al.* [59] also stated in their finding regarding the IR spectra of polyvinyl alcohol/tapioca starch bioplastics as shown in Figure 5.4. As can be observed in Figure 5.4(a), the adjacent peaks at 2150 and 2088 cm^{-1} indicated that there were the hydrogen bonds among the amylose chains, as well as between the amylose and the amylopectin. While in pure PVA (Figure 5.4(b)), the peak at 2162 cm^{-1} was assigned to the combination of O-H stretching vibration. The single peak in PVA showed that the hydrogen bonds are only formed among PVA chains. Besides, the C-H stretching of CH2 were detected at the peak at 3000-2800 cm^{-1} [60]. This might be happened due to the presence of water in the pure PVA sample, thus creates the broad O-H stretching peaks to overlay with the C-H stretching peak.

5.4.3 Differential Scanning Calorimetry

The thermal analysis or behaviour of polymers can be obtained using the differential scanning calorimetry (DSC) [61–63]. DSC can supply

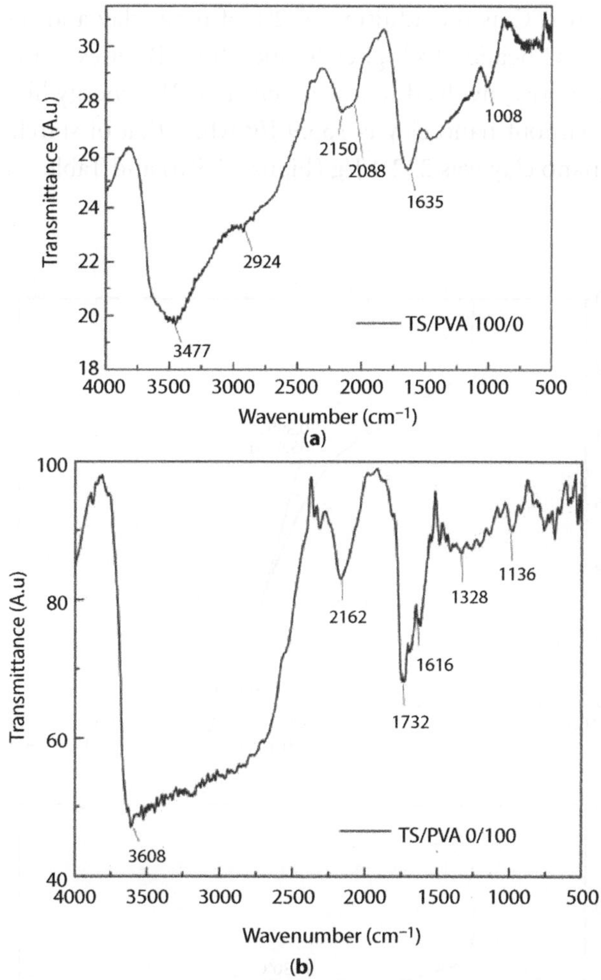

Figure 5.4 FTIR spectra of: (a) pure tapioca starch (TS) and (b) pure polyvinyl alcohol (PVA) bioplastic reference materials [59].

information on the susceptibility of different crystalline phases or arrangements to degradation through changes in the peak temperatures and shape of melt endotherms as well as the overall percentage crystallinity. As reported by Sadhu *et al.* [64] on the result of DSC analysis of nano clay filled starch-PVA blend showed that the pure starch/PVA blend at 50:50 ratio exhibited the peak temperature at 81.82°C with one exothermic peak starting at 50.39°C (Figure 5.5(a) and Table 5.1). The onset temperature was moved by 6°C as the addition of 2% of nano clay and the peak temperature by 2°C headed to higher temperature. Besides, a major amendment in the energy involved was also detected. The energy liberated in the pure blend without nano clay is 165.9 J/g while that of starch-PVA blend with 2% of nano clay was 222.4 J/g (Figure 5.5(b) and Table 5.1).

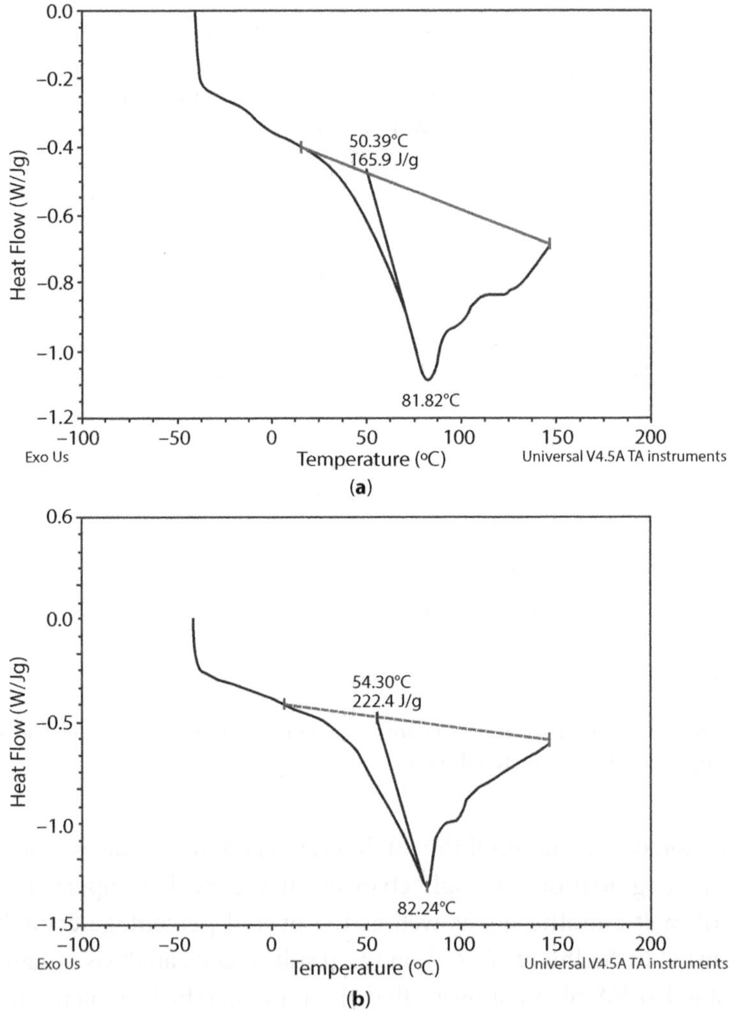

Figure 5.5 DSC curve of (a) starch-PVA blend at 50:50 ratio, and (b) starch-PVA blend at 50:50 ratio with 2% nano clay [64].

Table 5.1 DSC results of starch-PVA samples [64].

Sample	Onset temperature (°C)	Peak temperature (°C)	Energy involved (J/g)
PVA-starch blend	50.39	81.82	165.9
PVA-starch +2% nano	56.3	82.24	222.4

5.4.4 Thermogravimetric Analysis

Thermogravimetric analysis (TGA) is another technique used to study the thermal analysis of polymeric materials including PVA/starch blend. Apart from the DSC technique, TGA offers complimentary and supplementary characterization information. The amount and rate (velocity) of change in the mass of a sample as a function of temperature or time were measured by TGA in a controlled atmosphere. Primarily, the technique is used to investigate the compositional properties of materials as well as their thermal stabilities due to oxidation, decomposition, or loss of moisture.

For example, Zhijun Wu *et al.* [53] stated in their findings on the TGA of starch/polyvinyl alcohol/citric acid ternary (S/P/C) blend. The resulting TGA curves are shown in Figure 5.6 and the derivative thermogravimetric analysis (DTG) curves are shown in Figure 5.7. As can be seen from the thermo-gravimetric curves, the weight decreased along with increasing temperature while the DTGA curves displayed the maximum decomposition temperature (T max) of thermal decomposition [65]. The films exhibited multi-step thermal decomposition.

Due to the water evaporation, the initial thermal decomposition of the S/P/C1:1:0 was displayed from 90–105 °C, where the weight loss was 2.490 mg (10.24% of the total mass of the sample). Then, in the range of 200–320 °C, the main thermal decomposition was detected with the maximum decomposition rate around 310 °C, where the weight loss was 22.395 mg, (88.118% of the total mass of the sample. At this moment was the thermal decomposition of the hydrogen bonding between the starch and the PVA molecules. After the final thermal decomposition at 600 °C, the 52.149%, 57.025%, and 57.121% of the residuals were left for the $S/P/C_{3:1;0.08}$, $S/P/C_{3:3:0.08}$, and $S/P/C_{1:1:0}$ films, respectively.

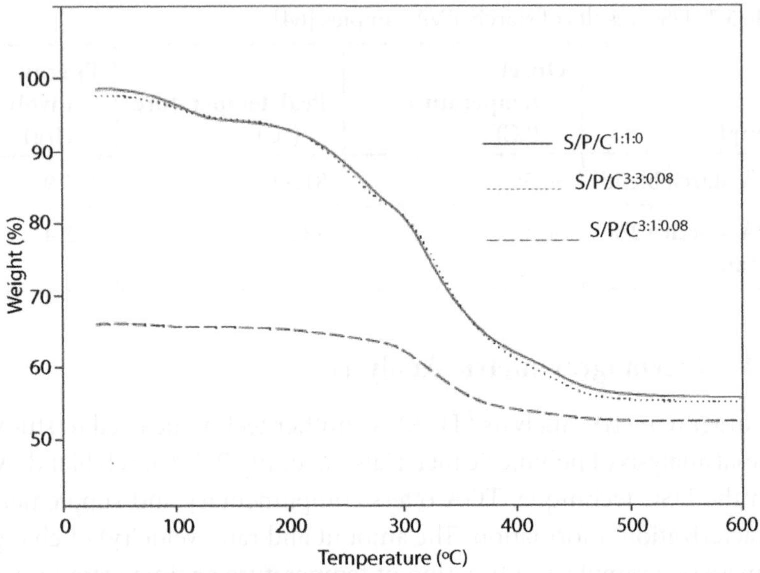

Figure 5.6 TGA thermograms of starch/polyvinyl alcohol/citric acid ternary blend functional food packaging films [53].

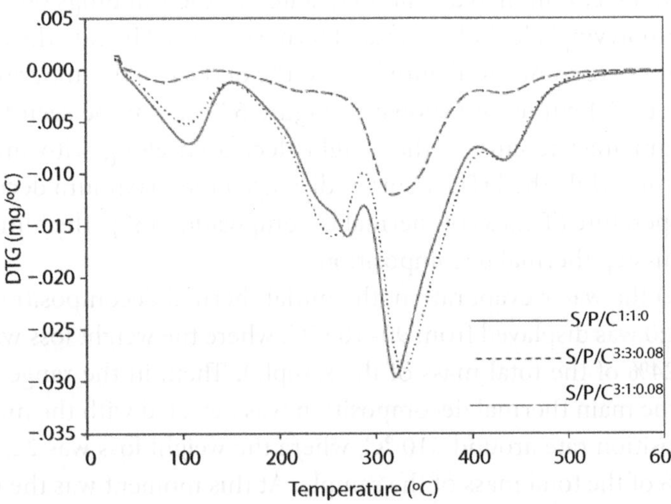

Figure 5.7 DTG thermograms of starch/polyvinyl alcohol/citric acid ternary blend functional food packaging films [53].

While for the $S/P/C_{3:3:0.08}$, the initial thermal decomposition was observed from 90–110 °C, due to water evaporation, where the weight loss was 1.7589 mg (7.387% of the total mass of the sample). Next, due to the modification of PVA by citric acid, the major thermal decomposition was seen around 200–320 °C with the maximum decomposition rate around 320 °C, where the weight loss was 21.234 mg (89.180% of the total mass of the sample). The addition of citric acid to the PVA led the starch and modified PVA molecules to form more new structure through hydrogen bonding [66]. Eventually, the cross-linking between the modified PVA and the starch became denser and the stability of the films was improved [67].

5.5 Applications

To this end, PVA/starch blend has become a favourable initiative for the preparation of biodegradable composites/nanocomposites, as the biocomposite is presenting sufficient hydrophilicity, mechanical and thermal properties, as well as biodegradable in the existence of applicably microorganisms. This biocomposite showed high potential use in packaging applications such as agricultural mulch film, water-soluble packaging and laundry bags for hospitals. Apart from that, other examples in packaging applications of PVA/starch materials include pesticides which are applied as a water spray, caustic cleaners or detergents which are dissolved during use, and process chemicals such as pigments, dyes or carbon black which are dissolved or dispersed in water.

5.6 Conclusion

PVA/starch biocomposites/bionanocomposites have presented a promising candidate to overcome the limitations of biodegradable blends. The synergistic effects shown by the PVA matrices and the starches were responsible for the improvement in overall performances of the PVA/starch blends. They have provided pronounce potential in mechanical properties as well as degradation behaviour. However, their tensile strength, elongation at break as well as their water resistance properties need to further improvement by the addition of compatibilizer or chemical modification. Thus, there are anxieties concerning poor stress-transfer efficiencies, poor interface quality between starches and polymer as well as high water permeability which should be apprehensive. Until now, current studies on the properties of biodegradable/starch blends have someway revealed some pronounced possibility to play a positive role in the certain application.

References

1. Tharanathan, R.N., Biodegradable films and composite coatings: Past, present and future. *Trends Food Sci. Technol.*, 14, 3, 71–78, 2003.
2. Achilias, D.S., Antonakou, E., Roupakias, C., Megalokonomos, P., Lappas, A., Recycling techniques of polyolefins from plastic waste. *Global NEST J.*, 10, 1, 114–122, 2008.
3. Karaduman, A., Pyrolysis of polystyrene plastic wastes with some organic compounds for enhancing styrene yield. *Energy Sources*, 24, 7, 667–674, 2002.
4. Andrady, A.L., Wavelength sensitivity in polymer photodegradation, in: *Polymer Analysis Polymer Physics, Advances in Polymer Science*, vol. 128, Springer, Berlin, Heidelberg, 1997.
5. Cataldo, F., On the ozone protection of polymers having non-conjugated unsaturation. *Polym. Degrad. Stab.*, 72, 287–296, 2001.
6. Andrady, A.L., Hamid, H.S., Torikai, A., Effects of climate change and UV-B on materials. *Photochem. Photobiol. Sci.*, 2, 1, 68–72, 2003.
7. Ismail, H. and Zaaba, N.F., Tensile properties, degradation behavior, and water absorption of sago starch plastic films. *J. Vinyl Addit. Technol.*, 18, 4, 235–240, 2012.
8. Zaaba, N.F., Ismail, H., Jaafar, M., A study of the degradation of compatibilized and uncompatibilized peanut shell powder/recycled polypropylene composites due to natural weathering. *J. Vinyl Addit. Technol.*, 23, 4, 290–297, 2017.
9. Müller, R.-J., Kleeberg, I., Deckwer, W.-D., Biodegradation of polyesters containing aromatic constituents. *J. Biotechnol.*, 86, 87–95, 2001.
10. Danjaji, I.D., Nawang, R., Ishiaku, U.S., Ismail, H., Mohd Ishak, Z.A.M., Degradation studies and moisture uptake of sago-starch-filled linear low-density polyethylene composites. *Polym. Test.*, 21, 1, 75–81, 2002.
11. Hayden, K.W., Jaimys, A., Russell, J.C., Ivanova, E.P., Plastic degradation and its environmental implications with special reference to poly(ethylene terephthalate). *Polymers*, 5, 1–18, 2013.

12. Yamada-Onodera, K., Mukumoto, H., Katsuyaya, Y., Saiganji, A., Tani, Y., Degradation of polyethylene by a fungus, Penicillium simplicissimum YK. *Polym. Degrad. Stab.*, 72, 323–327, 2001.
13. Zheng, Y., Yanful, E.K., Bassi, A.S., A review of plastic waste biodegradation. *Crit. Rev. Biotechnol.*, 25, 243–250, 2005.
14. Marqués-Calvo, M.S., Cerdà-Cuéllar, M., Kint, D.P.R., Bou, J.J., Muñoz-Guerra, S., Enzymatic and microbial biodegradability of poly(ethylene terephthalate) copolymers containing nitrated units. *Polym. Degrad. Stab.*, 91, 663–671, 2006.
15. Bonhomme, S., Cuer, A., Delort, A.-M., Lemaire, J., Sancelme, M., Scott, G., Environmental degradation of polyethylene. *Polym. Degrad. Stab.*, 81, 441–452, 2003.
16. Mohanty, A., Misra, M., Drzal, L., Sustainable bio-composites from renewable resources: Opportunities and challenges in the green materials world. *J. Polym. Environ.*, 10, 19–26, 2002.
17. Stevens, E. and Goldstein, N., How green are green plastics? *Biocycle*, 43, 42–45, 2002.
18. Mohanty, A.K., Misra, M., Hinrichsen, G., Biofibres, biodegradable polymers and biocomposites: An overview. *Macromol. Mater. Eng.*, 276, 1–24, 2000.
19. Xiong, Z., Dai, X.Y., Zhang, R.Y., Tang, Z.B., Na, H.N., Zhu, J., Preparation of biobased monofunctional compatibilizer from cardanol to fabricate polylactide/starch blends with superior tensile properties. *Ind. Eng. Chem. Res.*, 53, 10653–10659, 2014.
20. Ismail, H. and Zaaba, N.F., The mechanical properties, water resistance and degradation behaviour of silica-filled sago starch/PVA plastic films. *J. Elastomers Plast.*, 46, 1, 96–109, 2014.
21. Leja, K. and Lewandowicz, G., Polymer biodegradation and biodegradable polymers–a review. *Pol. J. Environ. Stud.*, 19, 2, 255–266, 2010.
22. Zhang, J.-F. and Sun, X., Mechanical properties of poly(lactic acid)/starch composites compatibilized by maleic anhydride. *Biomacromolecules*, 5, 1446–1451, 2004.
23. Nakashima, T., Xu, C., Bin, Y., Matsuo, M., Morphology and mechanical properties of poly(vinyl alcohol) and starch blends prepared by gelation/crystallization from solutions. *Colloid Polym. Sci.*, 279, 7, 646–654, 2001.
24. Huang, X. and Netravali, A., Biodegradable green composites made using bamboo micro/nano fibrils and chemically modified soy protein resin. *Compos. Sci. Technol.*, 69, 7, 1009–1015, 2009.
25. Witt, U., Müller, R.J., Deckwer, W.D., Biodegradation of polyester copolymers containing aromatic compounds. *J. Macromol. Sci. A*, 32, 4, 851–856, 1995.

26. Ishigaki, T., Kawagoshi, Y., Ike, M., Fujita, M., Biodegradation of polyvinyl-alcohol-starch blend plastic film. *World J. Microbiol. Biotechnol.*, 15, 3, 321–327, 1999.
27. Azahari, N.A., Othman, N., Ismail, H., Biodegradation studies of polyvinyl alcohol/corn starch blend films in solid and solution media. *J. Phys. Sci.*, 22, 2, 15–31, 2011.
28. Chiellini, E., Corti, A., Solaro, R., Biodegradation of poly(vinyl alcohol) based blown films under different environmental conditions. *Polym. Degrad. Stab.*, 64, 305, 1999.
29. Faria, F.O., Vercelheze, A.E.S., Mali, S., Physical properties of biodegradable films based on cassava starch, polyvinyl alcohol and montmorillonite. *Quím. Nova*, 35, 3, 487–492, 2012.
30. Guohua, Z., Ya, L., Cuilan, F., Min, Z., Caiqiong, Z., Zongdao, C., Water resistance, mechanical properties and biodegradability of methylated-cornstarch/poly(vinyl alcohol) blend film. *Polym. Degrad. Stab.*, 91, 4, 703–711, 2006.
31. Taghizadeh, M.T. and Sabouri, N., Thermal degradation behavior of polyvinyl alcohol/starch/carboxymethyl cellulose/clay nanocomposites. *Univers. J. Chem.*, 1, 2, 21–29, 2013.
32. Zaaba, N.F., Ismail, H., Jaafar, M., The effects of modifying peanut shell powder with polyvinyl alcohol on the properties of recycled polypropylene and peanut shell powder composites. *BioResources*, 9, 2, 2128–2142, 2014.
33. Zou, G.-X., Jin, P.-Q., Xin, L.-Z., Extruded starch/PVA composites: Water resistance, thermal properties and morphology. *J. Elastomers Plast.*, 40, 4, 303–316, 2008.
34. Xiong, H., Tang, S., Zou, P., Tang, H., Effect of nano-SiO2 on the performance of starch/polyvinyl alcohol blend films. *Carbohydr. Polym.*, 72, 3, 521–526, 2008.
35. Schmiele, M., Sampaio, U.M., Silva Clerici, M.T.P., Chapter 1-Basic principles: Composition and properties of starch, in: *Starches for Food Application*, M.T.P. Silva Clerici and M. Schmiele (Eds.), pp. 1–22, Academic Press, Brazil, 2019.
36. Yun, Y.H., Wee, Y.J., Byun, H.S., Yoon, S.D., Biodegradability of chemically modified starch (RS4)/PVA blend films: Part 2. *J. Polym. Environ.*, 16, 1, 12–18, 2008.
37. Kondo, T., Sawatari, C., Manley, R.S.J., Gray, D.G., Characterization of hydrogen bonding in cellulose-synthetic polymer blends systems with regioselectively substituted methycellulose. *Macromolecules*, 27, 1, 210–215, 1994.
38. Tang, S., Zou, P., Xiong, H., Tang, H., Effect of nano-SiO2 on the performance of starch/polyvinyl alcohol blend films. *Carbohydr. Polym.*, 72, 3, 521–526, 2007.

39. Yun, Y.-H., Wee, Y.-W., Byun, H.-S., Yoon, S.-D., Biodegradability of chemically modified starch (RS4)/PVA blend films: Part 2. *J. Polym. Environ.*, 16, 1, 12–18, 2008.
40. Wang, H., Sun, X., Seib, P., Strengthening blends of poly(lactic acid) and starch with methylenediphenyl diisocyanate. *J. Appl. Polym. Sci.*, 82, 7, 1761–1767, 2001.
41. Kalambur, S. and Rizvi, S.S.H., An overview of starch-based plastic blends from reactive extrusion. *J. Plast. Film Sheeting*, 22, 1, 39–58, 2006.
42. Mani, R., Tang, J., Bhattacharya, M., Synthesis and characterization of starch-graft-polycaprolactone as compatibilizer for starch/polycaprolactone blends. *Macromol. Rapid Commun.*, 19, 283–286, 1998.
43. Halimatudahliana, H., II and Nasir, M., The effect of various compatibilizers on mechanical properties of polystyrene/polypropylene blend. *Polym. Test.*, 21, 2, 163–170, 2002.
44. Mischnick, P. and Kuhn, G., Model studies methyl amyloses: Correlation between reaction conditions and primary structure. *Carbohydr. Res.*, 290, 199–207, 1996.
45. Liu, Z., Dong, Y., Men, H., Jiang, M., Tong, J., Zhou, J., Post-crosslinking modification of thermoplastic starch/PVA blend films by using sodium hexametaphosphate. *Carbohydr. Polym.*, 89, 473–477, 2012.
46. Bastioli, C., Bellotti, V., Giudice, L.D., Gilli, G., Mater-bi: Properties and biodegradability. *J. Environ. Polym. Degrad.*, 1, 181–191, 1993.
47. Iwanami, T. and Uemura, T., Properties and applications for packaging of starch/modifed PVA alloys properties and positive use of biodegradable new polymer "Mater-bi". *Polym. Preprints Japan*, 41, 2210–2212, 1992.
48. Tudorachi, C.N., Cascaval, M., Rusu, M., Pruteanu, M., Testing of polyvinyl alcohol and starch mixtures as biodegradable polymeric materials. *Polym. Test.*, 19, 785–799, 2000.
49. Orhan, Y., Hrenovic, J., Buyukgungor, H., Biodegradation of plastic compost bags under controlled soil conditions. *Acta Chim. Slov.*, 51, 3, 579–588, 2004.
50. Ismail, H. and Zaaba, N.F., Effect of additives on properties of polyvinyl alcohol (PVA)/tapioca starch biodegradable films. *Polym. Plast. Technol. Eng.*, 50, 12, 1214–1219, 2011.
51. Denis, M.P., Adriana, N.F., Marius, G., Ioana, C., Influence of storage conditions on starch/PVA films containing cellulose nanofibers. *Ind. Crops Prod.*, 70, 170–177, 2015.
52. Sadhu, S.D., Soni, S.G.V.A., Garg, M., Preparation of starch-poly vinyl alcohol (PVA) blend using potato and study of its mechanical properties. *Int. J. Pharm. Sci. Invent.*, 3, 3, 33–37, 2014.
53. Wu, J., Zhijun Wu, T., Li, Y., Lin, D., Xing, B., Yang, Y., Li, C., Yang, L., Zhang, L., Ma, R., Dai, J., Lv, X., Han, G., Preparation and application of starch/

polyvinyl alcohol/citric acid ternary blend antimicrobial functional food packaging films. *Polymers*, 9, 102, 1–19, 2017.

54. Xiong, H.-G., Tang, S.-W., Tang, H.-L., Zou, P., The structure and properties of a starch-based biodegradable film. *Carbohydr. Polym.*, 71, 2, 263–268, 2008.
55. Tang, S., Xiong, H., Zou, P., Tang, H., Effect of nano-SiO_2 on the performance of starch/polyvinyl alcohol blend films. *Carbohydr. Polym.*, 72, 521–526, 2008.
56. Das, K., Ray, D., Bandyopadhyay, N.R., Gupta, A., Sengupta, S., Sahoo, S., Mohanty, A., Misra, M., Preparation and characterization of cross-linked starch/poly(vinyl alcohol) green films with low moisture absorption. *Ind. Eng. Chem. Res.*, 49, 2176–2185, 2010.
57. Gulmine, J.V., Janissek, P.R., Heise, H.M., Akcelrud, L., Degradation profile of polyethylene after artificial accelerated weathering. *Polym. Degrad. Stab.*, 79, 385–397, 2003.
58. Edge, M., Infrared spectroscopy in analysis of polymer degradation, in: *Encyclopedia of Analytical Chemistry*, R.A. Meyers, (Ed.), John Wiley and Sons Ltd., Chichester, 2000.
59. Judawisastra, H., Sitohang, R.D.R., Marta, L., Mardiyati, Water absorption and its effect on the tensile properties of tapioca starch/polyvinyl alcohol bioplastics. *IOP Conf. Ser. Mater. Sci. Eng.*, 223, 012066, 1–10, 2017.
60. Shehap, A.M., Thermal and spectroscopic studies of polyvinyl alcohol/sodium carboxy methyl cellulose blends. *Egypt. J. Solids*, 31, 1, 75–91, 2008.
61. Olivares, N., Tiemblo, P., Gómez-Elvira, J.M., Physicochemical processes along the early stages of the thermal degradation of isotactic polypropylene I. Evolution of the γ relaxation under oxidative conditions. *Polym. Degrad. Stab.*, 65, 2, 297–302, 1999.
62. Rosa, D.S., Sarti, J., Mei, L.H., II, Filho, M.M., Silveira, S., Test method, a stuy of parameters interfering in oxidative induction time (OIT) results obtained by differential scanning calorimetry inpolyolefin. *Polym. Test.*, 19, 523–531, 2000.
63. Elvira, M., Tiemblo, P., Gómez-Elvira, J.M., Changes in the crystalline phase during the thermo-oxidation of a metallocene isotactic polypropylene. A DSC study. *Polym. Degrad. Stab.*, 83, 509–518, 2004.
64. Sadhu, S.D., Soni, A., Garg, M., Thermal studies of the starch and polyvinyl alcohol based film and its nano composites. *J. Nanomed. Nanotechnol.*, 7, 002, 1–5, 2015.
65. Rhim, J.W., Wang, L.F., Hong, S., II, Preparation and characterization of agar/AgNPs composite films with antimicrobial activity. *Food Hydrocoll.*, 33, 327–335, 2013.

66. Selke, S.E.M., Cutler, J.D., Hernandez, R.J., Plastics packaging: Properties, processing, applications and regulations, 2nd Edition, Hanser, Munich, Germany, 2004.
67. Gomez-Ordonez, E. and Ruperez, P., FTIR-ATR spectroscopy as a tool for polysaccharide identification in edible brown and red seaweeds. *Food Hydrocoll.*, 25, 1514–1520, 2011.

6

Polyvinyl Alcohol/Polylactic Acid-Based Biocomposites and Bionanocomposites

Ashitha Jose and Radhakrishnan E.K.*

School of Biosciences, Mahatma Gandhi University, Kottayam, Kerala

Abstract

The science of material engineering is currently evolving at a rapid pace in search of tailor-made polymers for specific applications in all sector varying from medicine to agriculture. Among the various polymers exploited till date, the polyvinyl alcohol (PVA) and poly lactic acid (PLA) have found major applications owing to its easy availability, cost effectiveness, versatile properties and so on. However, its application is limited by the native properties of the polymer. The incorporation of plasticizers and filler molecules in the form of nanomaterials and natural components are aimed at both enhancing the material characteristics and imparting specific activity such as antimicrobial potential to the developed bionanocomposites. In the food industry, PVA and PLA based polymer composites have applications to enhance the shelf life of the product and also act as sensors to determine the purity and safety of the packed food. On the other hand, in medicine, they find applications as various replacements in the body and also in the development of a range of medical devices. Tailor made bionanocomposites based on PVA and PLA can be obtained through various methods such as melt compounding, solvent based methods and electrospinning techniques and can effectively modulate the physical and mechanical properties of the formed polymer scaffold. For the evaluation of the property enhancement of the polymer composites, techniques including FTIR, microscopic techniques, thermal property analysis, barrier property analysis etc. are used. From the recent studies, addition of natural materials to the polymer matrix is also found to enhance the biodegradability issues raised by these polymers in addition to its property enhancement.

Keywords: Biocomposites, bionanocomposites, PVA, PLA, electrospinning

**Corresponding author*: radhakrishnanek@mgu.ac.in

Visakh P. M. and Olga B. Nazarenko (eds.) Polyvinyl Alcohol-Based Biocomposites and Bionanocomposites, (151–178) © 2023 Scrivener Publishing LLC

6.1 Introduction

The polymers, macromolecules having versatile structure and properties, have found a wide range of applications in all the aspects of life since long back. Their applications range from packaging to agriculture, medicine and finally the production of engineering parts. However, many of them are not environmentally benign both in terms of biodegradability and toxicity. This has led the research towards environmentally sustainable technology, which in turn has led to the development of the nature derived renewable materials offering both weight reduction and added functionality at a reasonable cost. Even though the natural fibers remain a widely accepted alternative to the artificial ones while considering the environmental impacts, their native properties alone cannot meet the physico-chemical requirements in terms of strength, elasticity, shear stress etc. This in turn has led to the development of the composite polymers, which are a blend of more than one polymer, a novel approach in the material chemistry aspects, which in turn has made the development of novel materials with tailor made properties at relatively low cost, an easy task at hand [1]. Recently the biopolymeric materials have been introduced as an alternative to the petrochemical polymers in all the field [2]. In another approach, the properties of the native polymers are altered through the addition of a wide range of fillers and plasticizers which can be exploited for the development of tailor-made properties of the polymer in question.

Polymers find a variety of applications in medicine ranging from the development of heart valves, blood vessels, urinary catheters, artificial skin and kidney to hemodialysis membranes. They have found applications in the nano based drug delivery systems, where, the biodegradable polymers are exploited. Recently, both the treatment and imaging techniques make use of PLGA (poly lactic-co-glycolic acid), PGA (poly glycolic acid) and PLA (poly lactic acid) which has been approved by the FDA for the medicinal usage. The polymers such as poly (ethylene terephthalate), polyurethanes, and microporous silicon rubber finds its application in the vascular prosthesis, while, polypropylene is used in cardiopulmonary bypass surgery. The semipermeable membranes of cellulose have been exploited in the hemodialysis. The mucoadhesive polymers are used in the ophthalmic and the buccal delivery of the drugs. Polyacrylic acid (PAAc), is a common bioadhesive drug delivery system. The controlled release systems such as the transdermal patches, microspheres, ocular impacts and contraceptive impacts also make use of a wide range of polymers. In the food sector, the polymers find a myriad of applications especially in the development of edible coatings and packaging materials. Edible polymers can either be made of carbohydrates, proteins, fibers or lipids and these coating materials are generally categorized as hydrocolloids, lipids and composites. They find application in different sectors of the food industry like the food packaging and detection of the toxic substances present in the food. The polymers are widely applied in the development of foamed food, snacks, micro and nano-encapsulated food, hydrogels, oligomers and prebiotics. They have found applications in the novel printed and electrospinned food as well. The commonly exploited edible polymers are starch, cellulose and its derivatives, casein, whey protein, alginate, gelatin, chitosan, collagen etc [3].

While considering its applications in other fields, polymer bonded permanent magnets find a quite significant application in the electronics sector - in both electrical and electronic devices. The magnetic composites are favored in the electrical machines due to its mechanical, magnetic and physical properties. Moreover, these fibers which are reinforced with polymers provide them both mechanical and magnetic functionality [4].

6.2 PVA Composites and Bionanocomposites

PVA, the first completely synthesized polymer is formed by the polymerization of numerous vinyl alcohol moieties. However, the lower abundance or rather the difficulty in isolation of the monomer makes PVA unique in terms that it cannot be obtained by polymerization. However, the desirable properties of the material added on to its wide acceptance across the globe, which in turn led to the production to 500000 tons of various grades of the material in just 50 years of its development [5]. The PVA is a water-soluble synthetic polymer having film forming, emulsifying and adhesive properties. Moreover, it is non-toxic and resistant to a variety of solvents including various oils. Its strength and flexibility remain an added advantage while considering the material for various applications. Its lower environmental impact - in terms of the solubility and biodegradability - has led to its wide fledged use. The solubility of PVA in water reduces with increase in the molecular weight which can lead to the formation of more inter and intra molecular hydrogen bonds [6].

PVA can be in either of the two forms, partially hydrolyzed or fully hydrolyzed. The variations in the solubility, flexibility, molecular weight adhesiveness and tensile strength of the PVA can be attributed to the variations in the length of both vinyl acetate polymer and the degree of hydrolysis [7]. PVA found its first application in the rayon textiles industry as a wrap sizing material along with its applications in the paper coating [8]. It is easily biodegradable and maintains its crystalline structure in water. It has found its applications in almost all the sectors ranging from industrial, medicinal to commercial. Other common applications of PVA include the synthesis of the biodegradable protective apparel, paper products, tube winding, carton sealing, thickening agent for latex paint, common household glue, and in other adhesive mixtures [7]. It has also found application as an ophthalmic demulcent, as an approved meat packaging product in transdermal patches, as a sustained release formulation in medicine, as a synthetic tear and in sustained release tablet formulations. The usage of PVA has surged over the time due to its wide applicability, which has resulted in larger release into the environment which in turn result in environmental issues if it remains untreated. PVA, not only enhance the Chemical Oxygen Demand (COD) but also the viscosity of the environmental water and thus result in the release of harmful metal ions from the sediments [9].

PVA finds its applications in almost all the aspects of life from food to medicine and finally to technology. In the food sector, it is exploited as a packaging material alone and in combination with other polymers and varied fillers. The starch capped silver nanoparticle incorporated PVA matrices have been developed as an efficient packaging material. These developed materials have superior characteristics in terms of the barrier, mechanical and biodegradation properties. Moreover, the silver nanoparticles synthesised using this system in the presence of sunlight has been found to have efficient antimicrobial activity as well [10, 11]. Similarly, the starch PVA composites having incorporated zinc oxide nanoparticles and phytochemical extract are able act as an intelligent p^{H} sensing wraps which can detect the food spoilage [12]. Modified PVA hydrogels, a three-dimensional polymer network has been employed commonly in medicine due to its ability to absorb considerably large amount of water or biological fluids. PVA based drug delivery systems have become common. It can be obtained through the conjugation of PVA with either micro or nanoparticles including the paclitaxel loaded poly(vinyl alcohol)-graft-poly (lactide-co-glycolide) (PVA-g_PLGA NP) for the treatment of restenosis.

PVA cross linked systems have been used for the encapsulation of theophylline. The low absorption of PVA has led to its emergence as a non-absorbing carrier matrix as it can be used for the nasal delivery of acyclovir. The potential of PVA is being exploited in developing novel magnetic drug targeting systems to overcome the harmful side effects of the chemotherapy. The skin regeneration ability of the PVA/PVP hydrogels are being exploited towards its development as wound and burn dressings. It is also being recognized by the US FDA as an ophthalmic demulcent which is currently available in the soft contact lenses [13].

6.3 Poly Lactic Acid (PLA) Composites and Bionanocomposites

Poly lactic acid is a thermoplastic aliphatic polyester which can be derived from renewable sources such as corn starch, sugarcane and tapioca roots. PLA remains one of the most promising polymers considering its properties such as biocompatibility, biodegradability, ready availability and low cost, in addition to its outstanding processing performance. However, the brittleness and poor thermal stability remains the limiting factors [14]. The copolymerization of the PLA or the introduction of plasticizer has thus become a suitable approach to tune its characteristics. The modification of the PLA can be bought about through the employment of various techniques such as copolymerization, crosslinking and blend formation which in turn can lead to the development of tailor-made material. However, it has been proposed that better toughening results could be obtained through the incorporation of nanoparticles as fillers [14]. The mechanical properties of composites can be varied significantly with the type and dispersion of fillers employed in the matrix. According to Hoidy *et al.* [15] the mechanical properties of the PLA could be improved by the incorporation of MMT clay. The PLA films have found its application in both medicine and packaging and is being considered as an alternative to the synthetic packaging materials. The biocompatibility and the biodegradability of the PLA films in addition to its superior tensile properties have brought it to the forefront of the human interface in spite of its higher hydrophobicity. However, it lacks impact toughness as evident from the low elongation at break, which in turn limits its applications. This can be tackled through the wettability improvement as brought about through the blending process which utilizes the compounds such as chitosan, collagen, elastin etc. Recently, the PLA blends have come up with properties such as shape memory and morphology which is non-existent in the parent polymers. Furthermore, these blends have found its application in the development of scaffolds, fibres and films [16].

Removal of the temperature or the solvent during the blend formation can lead to the separation of the phases which initially was miscible and the degree of this separation can affect the morphology of the final product. The spinodal decomposition and the nucleation – growth are the two common processes involved in the development of the blends where the former technique can be exploited towards the development of tailor-made products for various commercial application. The industrial production of PLA occurs through 2 major routes namely the direct polycondensation of the lactic acid and the ring opening polymerization (ROP) via the cyclic dimer. The direct polycondensation mechanism involves the condensation which leads to the removal of the ester and the use of solvent under high vacuum and temperature. In this method, the production of low to intermediate molecular weight PLA is possible. The removal of water and other impurities remain a limiting factor. The chain extension method can be exploited to overcome the deleterious effects of the process which however will result in the development of the PLA polymer whose properties might be altered with the applied procedure.

The higher molecular weight polymer can be obtained through the ROP of lactide which exploits the stannous octotate as an industrial catalyst. This multi-step process does not involve the use of any solvents. The aqueous lactic acid undergoes condensation to form a low molecular weight prepolymer. It is then converted into a mixture of lactide stereoisomers in the presence of tin catalyst which will enhance the selectivity and rate of the intramolecular cyclization. Then, distillation is employed to remove the impurities wherein the meso-lactide is separated and combined with low D-lactide fraction to generate larger portfolio of PLA grades. The synthesis of higher molecular mass PLA includes the tin catalyst by the ROP of lactide in the melt which completely eliminates the use of costly and environmentally unfriendly solvents. The industrial ROP of the lactide can result in the generation of PLA copolymers of different isomer ratios. The polymerization of lactide by the reactive extrusion (REX) can be used to obtain the PLA continuously in larger quantities and lower costs.

6.4 The Role of Plasticizers and Fillers in Composite Development

Plasticizers, low molecular weight compounds have been exploited in the development of composite materials. They can occupy the intermolecular spaces of the polymeric chains, thus altering the three-dimensional organization thereby enhancing the molecular mobility and free volume [17]. The common plasticizers include the polyols and saccharides in hydrophilic polymers while, glycerol finds its application in hydrocolloid-based films wherein it can enhance the flexibility by reducing the brittleness [18].

The fillers can be any smaller molecules which can have impact on the physicochemical properties as well as the functioning of the matrix under consideration. Herbal constituents, small molecules, peptides, drugs and even nanoparticle can act as a filler. The alterations in the properties can be optimized based on the filler used and its interaction with the matrix. It can have significant effect on the barrier properties of the developed material. The encapsulation technology is exploited in enclosing the active substances inside the polymer of interest. This improves the stability of the active ingredient. The nanofibers possessing high surface area are used in this aspect. The release of the filler molecules from the film matrix can be determined based on the peak area as obtained from the head space GCMS during various time intervals. The PLA films loaded with carvacrol has shown a dose dependent initial burst followed by a sustained release of the filler molecules over the time. The variations in the release pattern remains subjected to the concentration of the filler, which affected the morphology and size of the fibres [19]. The addition of rice bran oil and soybean oil to the PLA matrix was able to bring about a trend similar to that of plasticizer-polymer system. Even though the oil fractions remain as a separate phase at lower temperature, certain compatibility do exist between the 2 phases as complete de-mixing is not possible. Moreover, the addition of rice bran oil in smaller percentage was able to develop PLA materials containing 'alpha' phase [20].

The increase in the essential oil concentrations in the film is said to result in highly interconnected fibers using the electrospinning procedure.

This could be attributed to the variations caused in the thermal properties of the composite fiber. However, the surface topography of the formed fiber is dependent on the polymer used. To elaborate, the addition of different amounts of carvacrol into the zein solution resulted in the film having smooth morphology in the absence of any pores and pits. However, when the same was added into the PLA matrix, a textured topography was obtained. Increased surface roughness was confirmed along with the formation of few pits when the concentration of the filler was increased from 5 to 10%, deeper pits and wrinkles were resulted in fibers loaded with 20%. This wrinkled surface could be attributed to the shrinkage of the polymer jet [19].

6.5 Methods Employed in the Development of Structured Polymers

6.5.1 Melt Compounding

In melt compounding, the dry mixing of all or some components is done using high speed mixers followed by the compounding of the same in any of the melt-blending equipment's including internal mixers, single/twin screw extruder or buss kneader. The type of the equipment, the nature of the dispersed phase and the desired end use of the product can determine the processing conditions. The major advantage of this method, the non-usage of any organic solvent, make it a highly preferable one in terms of sustainable development. Moreover, the upgradation of this method to large-scale and industrial scale is comparatively easy [21].

6.5.2 Solvent-Based Methods

The solvent casting methods find its importance in the early phases of research especially in smaller laboratory scale. The lower quantity of the active pharmaceutical ingredients (API) has led to the probing of this method in the medical sector [21].

6.5.3 Electrospinning

Electrospinning is an efficient method explored for the development of structured polymer fibers having diameter range from a few micrometers to tens of nanometer. The so developed fibers find its use in terms of filter application, nanotube development, biomedical applications and as reinforcing materials [22].

The procedure involves application of a strong electrostatic field to the capillary tube which in turn is connected with the polymer solution reservoir leading to the formation of a pendant droplet of the corresponding solution at its tip. When the applied voltage is high enough to overcome the surface tension, a fine charged jet is ejected, which moves towards the ground plate which act as the counter electrode. The entanglements and the viscosity of the polymer solution are responsible for the maintenance of this jet. The solvent phase evaporates soon after the jet formation resulting in the deposition of a thin polymer fibre upon the substrate which is located on the counter electrode. This deposition rate can extend to several meters per second. Furthermore, the acceleration towards the counter electrode can result in the strong elongation which increases the surface within milliseconds. This in turn helps in overcoming the phase boundaries. The processing parameters chosen can lead to the formation of tailor-made diameters of the fibre. The internal morphology of the developed fiber plays a crucial role in determining the applications. Porous fiber, which can be generated by ensuring the phase separation to occur during the electrospinning process, find its application in filtration and preparation of nanotubes. The pore formation can be attributed to the solvent rich regions prior to the phase transformation. The lower pore formation as seen in solvents with lower vapor pressure confirms this theory. An electrostatically driven jet of polymer is used to obtain submicron polymer particles, fibers or interconnected porous fiber meshes in electrospinning. It remains a relatively simple technique in generating porous structures. The structure of the obtained polymer is dependent on the concentration of the solution studied along with its molecular weight [23].

6.5.3.1 *Melt Electrospinning*

The major drawback of solution electrospinning in its biomedical application is the presence of residual toxic solvents within the prepared nanomeshes which can be answered through the development of melt electrospinning method which exploits the polymer melts instead. It is the melt electrospinning writing, a novel approach in the development of electrospun fibers of highly porous scaffolds, generated through the precise deposition of the micron scaled fibers in a consecutive fashion, which has recently flourished in terms of its applications. The technology as such can be considered as a process which moves through the polymer preparation followed by the jet generation and finally the fiber collection. The nonwoven mats synthesized through the electrospinning methods finds its application in purification and removal of various adsorbents owing to its high surface area and permeability to water [24, 25].

6.5.3.2 Near Feld Electrospinning (NFES)

The process of direct written nanostructure or orderly arranged nanoarrays in a controlled and continuous manner is termed as the near field electrospinning. Here, the electrode-collector distance can be as short as 500μm to 3mm while the applied voltage varies in between 100-600V. The advancements with NFES is its ability to position the individual fibers with a great precision which can be achieved through the minimalization of the needle-collector distance. Further, the technique can result in the development of fibers with improved morphology. Moreover, in NFES, it is the initial polymer jet diameter which determines the final nanofiber diameter instead of the bending instability [26].

6.5.3.3 Electrohydrodynamic (EHD)

Electrohydrodynamic (EHD), is another technology wherein the direct writing of the structures, either in two or three dimensions is possible. The electric potential applied between the nozzle and the substrate is much smaller when compared to that used in NFES resulting in a pulsating mode (nanodripping) rather than a stable jet mode. Even though, the droplet diameter has an inverse relationship with the applied voltage, the diameter remains constant after a certain threshold concentration [26]. The characteristics of the nanospun fiber can be determined based on various parameters such as the solution parameter, the process parameter and ambient external conditions. The solution concentration plays a crucial role in determining the fiber dimeter. The lower polymer concentrations can result in lower diameter of the fiber. While too low concentration (decreased to the entanglement concentration) can result in the production of beaded fibers, and too high concentration can result in the development of micro-ribbons having helix-shape. While, the increment in the feed rate can result in the lower charge density, higher charge density is responsible for the secondary bending instabilities in the fiber resulting in fibers with lower diameter. Similarly, increase in fiber diameter can be obtained through increasing the feed rate. However, too high flow rate can result in fibers with beads, as the system do not provide enough conditions for the solvent evaporation.

While a voltage beyond the threshold is essential for the electrospinning procedure. Higher voltage can result in the low diameter fibers owing to the higher electrostatic force of the solution which in turn can favor the stretching of the jet. The distance between he tip and the collector can determine the diameter and the morphology of the fiber. The minimum distance should be applied to ensure the solvent evaporation while longer distances can result in the thinner fibers. However, too small or large distance can result in the development of beads. The higher surface area to volume ratio of the nanofibers can lead to high absorption rate and thus find its application in various fields of study. It can also lead to the development of nanowoven structures having high conductivity finding its application in fuel cells and batteries. The limitations of the electrospinning include limitation in the number of polymers used, cost and limited range of performance application [26].

6.5.3.4 Coelectrospinning

The development of continuous fibers which encase materials (nanoparticles) within the polymer sleeve can be obtained through the co electrospinning process. The core shell polymer can be developed using a 2 stage approach wherein the core polymer is electrospun in the primary phase followed by the deposition of the shell polymer through coating in the second stage which can lead to the development of Core-Shell polymer. The advancement of this technique can be visualized in the single stage process of co electrospinning. Here, the core polymer is issued from the inner tube through the coannular nozzle while maintaining the annular coflow of the shell polymer solution. The technique requires a polymer solution in the shell along with a polymer or a nonpolymeric solution in the inner core [27].

The process can be deemed success when both the solutions used are sufficiently viscous and spinnable. Moreover, these solutions under consideration should not be miscible. While handling the miscible solutions for the process, the appropriate control over the system could not be guaranteed which in turn can lead to the mutual diffusion of the solution in the Taylor cone during the stretching process. The recent advancements in the techniques has enabled the system to coeletrospin the blend of 2 polymers through a single nozzle which results in core/shell fiber development. The chosen polymer pair can result in the development of the hollow polymeric fibers. These approaches are widely exploited in the development of carbon, ceramic or metallic tubes [28]. The formation and stabilization of microtubes can be affected by two mechanisms, the fast evaporation of the shell solvent and the contact with a nonsolvent, which can be attributed as a dry/wet electrospinning process. To elaborate, the outer surface the shell experiences a dry spinning process through its exposure to the surrounding air, leading to the diffusion of the solvent and its evaporation whereas the inner shell undergoes a wet spinning process. Here, as soon as the core solution starts to penetrate the shell, precipitation occurs immediately in the inner surface of the shell. The solvent systems used must have a higher affinity, the precipitation due to the inflow of the nonsolvent into the shell remains a quick process. Thus, the core remains wet over a longer period of time after the solidification of the shell. This fast/slow evaporation is essential in obtaining hollow tube structures. The evaporation of the core solution starts with a nucleation of bubbles, which then will grow to reach the inner diameter of the microtube. The growth of the bubbles occurs longitudinally as the tube wall confines the radial area. The multiple nucleation points which might occur due to the presence of the pores, holes, local thinning of the shell or other irregularities will recede independently in both directions to merge together. The fiber diameter and its aspect – hollow/solid-can be determined based on the flow ratio of both the solutions under consideration [28].

6.6 Techniques for Analyzing the Biocomposites and Bionanocomposites

6.6.1 FTIR

Fourier transform infrared spectroscopy is a widely accepted method to study the molecular structure of a compound. The environment and conformation of the macromolecules at molecular level can interfere with the width and intensity of the peaks. Also, the intermolecular interactions among the individual components can bring about changes in the obtained spectrum [29]. In the FTIR spectral analysis of PVA, the absorbance at 3700 – 3000 cm^{-1} is indicative of the stretching vibrations of OH group. The aliphatic CH stretching of the alkyl groups results in vibration bands at 2902cm^{-1} and 2938cm^{-1}. The CH bending vibrations can result in peaks at 1375cm^{-1} and 1365 cm^{-1}. The H-O-H bending of water can be confirmed by peaks at 1653 cm^{-1} and 1644 cm^{-1}. A peat at 1057 cm^{-1} in cellulose can be attributed to the C-O-C pyranose stretching vibration in aliphatic primary and secondary alcohols of cellulose [1]. In case of PLA, the stretching of the OH groups could be attributed to peaks at 3495cm^{-1}, 1759 related to the C=O stretching of carbonyl, 1184 and 1088 by C-O-C bending vibration, 1047 cue to the C-CH3 vibration. Bands at 1455 and 1384 could be attributed to the CH deformation from CH2, 870 and 757 defines the C-C stretching vibrations. The peaks at 862 and 755 due to the amorphous crystalline structure of the PLA fiber. The peak shifts and variations in the intensities could determine the interactions among the individual components [19]. The shift in the peak positioning could be attributed to the interactions between the fillers and the matrix material. To elaborate further the shift in the NH stretching of the pure zein fiber and the carvacrol loading as reported by Altan *et al.* [19] could be due to the interactions among the individual constituents, the phenols of the carvacrol and the zein. Similar changes have been reported during the interactions of gallic acid to the zein fibers.

6.6.2 Thermal Properties of Films

Differential scanning calorimetry (DSC) and thermogravimetric analysis (TGA), are perfomed to analyze the thermal stability of the polymer matrix. In TGA, the mass of the sample is analyzed over time with change in temperature. The weight change can be efficient in determining the volatile fraction present. This technique can be used to estimate the amount of filler employed in the final developed product. Meanwhile, DSC measures the heat flow into and out of the sample as a function of temperature or time. It measures the heat flux involved while varying the temperature. It is commonly employed to understand the physical changes such as the glass transition temperature.

The region between 30°C-100°C in the DSC curve can be attributed to the volatilization of water from the films while the region between 180°C-390°C to the depolymerization of the film components. T_{max} is the temperature corresponding to the DSC curves and an endothermic T_{max} peak at 221°C is due to the PVA. Large and sharper peaks can be obtained with an increment in the PVA concentration. However, this endothermic peak was softened with the addition of the plasticizer, glycerol. This could be due to the interaction of the PVA with glycerol resulting in corresponding film to be thermally stable. The thermal degradation of the cellulose occurs between 225 and 325°C [1].

In case of cellulose, weight loss at 100°C is due to the loss of water, at 230°C is due to the degradation process including depolymerization and decomposition of glycosyl groups leading to the formation of residues. Finally, the decomposition at 360°C is due to the oxidation and breakdown of char residues where in gaseous products of low molecular weight are formed. The residue at 850°C, the sulphate groups, remain as a flame retardant [30]. The neat PLA fibres have two stages of weight loss, initial (2.3%) below 80°C due to the solvent evaporation in case of electrospun fibres, and the major weight loss occurred between 220-355°C, with a degradation peak at 346°C. This could be attributed to the chain scission. The incorporation of fillers can result in varied TGA. In case of carvacrol loaded PLA, the initial weight loss could occur due to the evaporation of the filler molecule. And a shift in the degradation peak could be seen to be affected by the amount of filler added. However, the thermal stability of the complex system was found to be improved on the addition of filler which enhanced the hydrogen bonding of the system [19].

6.6.3 Scanning Electron Microscopy

The surface and lateral view of the prepared films can be obtained using SEM imaging techniques. The softening of the surface of the thin films, making them to be free of pores with rise in the PVA concentration can be confirmed through the imaging techniques [1]. The surface morphology of the electrospun fibers when observed under the FE-SEM can notify on the mode of evaporation of the solvent system. The rapid evaporation can result in the ribbon like structure, resulted from the collapsing of the fibres. The contact between the diametrically opposite parts of the fibre skin can lead to the formation of tubes at the edges of the ribbon as indicated in the

FESEM images. The change in the fibre diameter distribution as resulted from the addition of the filler can be determined using the same technique. This can be resulted from the varying charge density of the filler molecules. Higher charge density can result in better stretching forces that result in the self-repulsion of the jet which in turn reduces the fibre diameter. Furthermore, the reduced viscosity of the solution can result in thinner fibre formation as a result of the higher jet stretching.

6.6.4 TEM

Transmission electron microscopy is a technique used to determine the size of a molecule to size scaling a single atom. The outcome of the TEM imaging can be determined on a various factor such as the imaging type (dark/bright field) and magnification which in turn will result in the resolution, particle background and boundary determination and the number of particles per image. The technique remains unique when compared to other visualizing tools that it is the only one that allows for the direct/real space visualization of the samples [31].

6.6.5 Barrier Properties

6.6.5.1 Light Barrier Properties and Transparency

The oxidation of the packed food can bring about rancidity, undesired odor and flavor along with lower shelf life. This can be accelerated by the UV rays. Here comes the importance of the UV barrier properties of the material which can reduce the rate of oxidation. The oxidation of the lipids occurs when subjected to UV radiation at a range of 200-280. The plasticizer, glycerol is able to block the UV transmittance to a greater extend [1, 32, 33].

6.6.5.2 Oxygen Barrier Properties

Oxygen permeation analysers are used to determine the oxygen transmission rate which can be affected by the film thickness. The oxygen permeability can determine the use of the developed material in any aspect. Hence, the native barrier properties are modified accordingly by the addition of fillers/plasticizers along with the development of the composite material for the same. The addition of polyhydroxybutyrate into the PLA film matrix has resulted in the improved oxygen barrier property. The intrinsic high crystalline nature of the PHB resulted in the formation of crystalline order in the PLA matrix. The increase in the chain mobility of the matrix and the free volume through the blend formation can lead to further increment in the OTR as induced by the plasticizers. The composite films prepared with PVA and rice boiled starch, having *in situ* generated silver nanoparticles in the matrix is found to be efficient in preventing the entry of the microorganisms, ensuring its higher barrier property [34].

6.6.5.3 Water Vapour Barrier Property

The water absorption of the packaging material must be able to maintain the moisture content of the substance. The water vapour permeability can be analysed with the help of a desiccant method. The interaction of the material with both surface and atmospheric water can determine its application. While a perfect wound dressing material must be able to absorb higher amount water, the food application requires the water vapour penetration to be the least. The starch PVA composite films incorporated with zinc oxide nanoparticles and phytochemicals are found to be efficient in acting as a water barrier. To address the poor water vapour and oxygen barrier properties of PLA films, Goh *et al.* [35] has come up with a sandwich-architecture film of PLA where the impermeable reduced graphene oxide act as the core barrier and the commercial PLA films acted as a protective encapsulation. The excellent barrier property of these prepared films can be attributed to the hydrophobicity of the graphene oxide of the core barrier and the compact lamellar microstructure.

Ethanol has been identified as a potent food stimulant in identifying the overall migration of the non-volatile substances of the film that could be transferred to the packed food material and should be lower than 60mg/kg. The polarity and solubility remain the controlling factors in determining the rate of migration along with the chemical interactions among the individual components of the matrix. However, the plasticizing effects of the additives can result in the higher mobility of the low molecular mass compounds through the polymer chains resulting in an increased mobilization of the components.

6.7 Application of Polymers in the Food Industry

Any material which can prolong the shelf life or ensure the safety of the packed food by changing the existing packaging conditions can be defined as a packaging material. Packaging materials play an extremely important role in the food industry owing to their ability to protect the contents from mechanical and chemical damage and to a greater extend from microbial spoilage. The past trend of petrochemical based packaging materials has been replaced widely by biodegradable ones considering its positive impact on the environment. While choosing the apt packaging material, the mechanical properties of the material must ensure its integrity during the transport, handling and storage of the material [1]. The nanocomposite food packaging is a class of active packaging material wherein metal nanocomposites are incorporated to bring about the antimicrobial properties. The functional characterization of the prepared nanocomposite materials can help in determining its efficacy as a active packaging system. This can be confirmed through the barrier properties and the overall migration into the food or other stimulants, antioxidant and antimicrobial potential along with the disintegration and decomposition of the material [34, 36].

The novel packaging materials can be considered as an active packaging wherein the attenuated properties can enhance the shelf life. The material might be modified to possess antimicrobial, antioxidant, barrier properties, or can even be active packings which can visualize the food degradation subjected to the change in pH, hype in microbial growth or changes in rancidity of the packaging material. The individual properties possessed by the material are determined by the interactions among the polymers exploited and the fillers and plasticizers used. The development of biosensor-based packaging and barcoding can ensure the food safety. The intelligent pH sensing wraps developed using PVA/starch complex having incorporated zinc oxide nanoparticles and phytochemical constituents is a novel type of active packaging. Among the various packaging materials available, the antimicrobial packaging is a common application. Here, the matrix of the material is incorporated with antimicrobial components which can either be a nano or micro material or an herbal composition. The *in vitro* susceptibility of the bacteria to the active components present in the packaging material determines the antimicrobial potential of the film. The release pattern of the incorporated antimicrobial components is a determining factor of the food quality and safety during the packaging. A slower release rate allows the bloom in microorganism before the release of the compounds. However, a fast release could lead to the diffusion of the antimicrobial agents into the food preventing its prevalence on the surface and thus tamper with the shelf life. Thus, sustained release of the components coupled with an initial burst is essential to maintain the integrity of the developed antimicrobial packaging [19]. This can be assessed using either the disc diffusion method of the broth dilution methods. PVA/rice boiled starch matrix containing silver nanoparticles is an effective antimicrobial packaging material for the preservation of meat samples. Here the silver nanoparticles are responsible for the antibacterial potential of the developed membranes.

The antioxidant property, the ability to neutralize the reactive species present, can determine the health of a living form. The modified packaging materials composed of antioxidant materials are aimed at improving the shelf life of the packed material. The antioxidant potential of the developed films can be determined using any of the standard procedures including the DPPH assay. However, the release of the individual components need not be quick for a visual appreciation in a short time. The hydrophobic molecules find potent antimicrobial application as they get inserted easily between the fatty acid chains of the lipid bilayer of the membrane. This in turn could affect the cell size [34].

Considering its abundance and natural impact, cellulose have been exploited widely as a packaging material. Eventhough, considered as an almost inexhaustible source, its poor water solubility and mechanical properties has limited its application as a stand-alone molecule. Hence, the importance of composite films developed using water soluble polymers such as PVA wherein, the cellulose is exploited as a filler molecule. Here, the mechanical properties of the composite blends can be altered with ease [1, 2].

6.8 Application of Polymers in Medicine

The polymers find a wide range of applications in medicine. They are being exploited as the drug carriers for specific targeted delivery as well the sustainable release of the drug. The ophthalmic composition to the surgical interventions includes a wide range of polymers. Hydrogels are defined as gels containing water but not soluble in. PVA hydrogels have found its application as an artificial vascular material replacement considering its low toxicity, excellent mechanical properties comparable to the vascular system and biocompatibility. The freeze thaw method employed in the PVA hydrogel preparation can overcome the issue of turbidity. The method employs heating a mixture of PVA/water/DMSO (dimethyl sulfoxide), agitating under nitrogen current and then cooling the solution at temperature as low as -10°C for 10 hrs. This leads to the increased crystallization and cross-linking of the PVA molecules. The so formed hydrogels show enhanced properties in terms of mechanical strength, water content and transparency [37].

However, the physical cross linking of PVA using DMSO and DMF is found to be more efficient than the traditional freeze thaw method [38]. A water content of 92% was obtained on developing PVA hydrogels reinforced with cellulose nanocrystals which was cross linked by glutaraldehyde. The nanoparticle incorporation did not affect the swelling and thermal stability of the hydrogels. However, the compressive strength was improved and the creep elasticity reduced and the strain recovery improved in the modified hydrogels. Initially, PLA was exploited minimally in medicine which included the implant devices and tissue scaffolds along with internal sutures where the cost, limited molecular weight and availability remained the limiting factors. The novel techniques leading to the economic production of high molecular weight PLA has now led to its wide spread usage in medicine. The most important factors in the treatment of persistent infections is the requirement of higher local drug concentration along with the sustained release of the drug. The development of electrospun PLA fibers containing both uncoated and encapsulated chlorhexidine particles was found to be of use in the treatment of persistent infections of both dentistry and medicine. The fabrication of the chlorhexidine particles was mediated through the precipitation of the chlorhexidine diacetate with calcium chloride which was then encapsulated using the layer by layer approach.

These developed particles had a chlorhexidine content as high as 90% and resulted in a bead in the string structure upon electrospinning. This method was found to be efficient in ensuring sustained release over a period of 650 h. The release of the particles was mediated through a polyelectrolyte multilayer encapsulation as confirmed through the SEM and confocal imaging techniques. The non-toxic nature of the fibers along with its potent antibacterial activity can project its application as a sustained release material. The material was able to show a burst release in water wherein over 60% of the medicine was released in the first day followed by a sustained release in the upcoming days. Following the chlorhexidine release, collapsed spheres were observed in the fibers. The porous surface thus generated revealed the initial morphology can lead to the inference that the PLA layer has been penetrated. The usage of fresh solutions for the electrospinning was able to avoid the issues related to the dissolution of the chlorhexidine. Higher sustained release was confirmed through the pre-encapsulation of the chlorhexidine particles with the polyelectrolyte multilayers.

The introduction of carvacrol to the PLA matrix during electrospinning led to the formation of an aromatic ring band in the ATR region of 812cm^{-1} which confirmed the successful inclusion of the essential oil into the PLA matrix. Furthermore, a higher intensity of the band at higher loading (28% instead of 14%) also confirmed the interactions among the matrix and the essential oil. The low molecular weight of the compound made it to act as a plasticizer reducing the intermolecular forces of the polymer chain which in turn improves the flexibility and extensibility of the polymer matrix. Moreover, the enhancement to the surface interactions bought about by the carvacrol, hinders the slipping of the fibres and thus improved the elastic modulus. While considering the thermal characteristics, it was clear that the carvacrol led to the rubbery behaviour of the PLA instead of its glassy nature by acting as a plasticizer as similar to various other essential oils studied. The release study of the fraction confirmed a bust stage followed by a slower release as required for a sustained release system [39].

6.9 Biodegradability of PVA

The biologically decomposable polymers originate from petroleum based synthetic materials which decompose naturally under aerobic or anaerobic conditions. The drastic increment in the PVA utilization has led to the raise of substance in the water bodies across the globe after being used in the textile and paper industries. Its foam forming ability prevents the recovery of oxygen in the so polluted water bodies. Microorganisms ubiquitous to septic systems, landfills, composts and soil have been found to be able to degrade the PVA enzymatically. The oxidases and hydrolases when combined can convert the PVA into acetic acid. However, the rate of biodegradation is dependent on the percent hydrolysis and the solubility of the molecule. The blending of the PLA can alter the degradation profile of the composite eventhough the degradation profile remains identical to that of the parent polymer, the rate and the level depends on the composition and morphology of the individual components [16].

The first report on the biodegradation of PVA was by *Fusarium lini* B. Even though most of the PVA degrading bacteria are gram negative such as *Pseudomonads* and *Sphingomonads*, some gram-positive bacteria are also capable of degrading the same. Also, Penicillium sp. is an eukaryotic degrader of the PVA. While *Geotrichum fermentans* is a vinyl alcohol chain degrader fungus, lignolytic fungi is a nonspecific degrader. It is extensively accepted that the biodegradation of PVA occurs through a 2-step process. In the first step, the extracellular PVA oxidase secreted by the organism yields oxidized PVA, which undergo a spontaneous depolymerization followed by an enzymatic hydrolysis. While in case of gram-negative bacteria, the PVA is taken up into the periplasmic space where it is oxidized by PVA dehydrogenase coupled to the respiratory chain [40]. While analyzing the effect of UV irradiation and chlorination on the degradation on PVA, it was found that neither of them alone was capable of bringing about degradation. However, a combination of both was capable of degrading the PVA films quite faster. This occurred through the generation of OH and Cl radicals, among which, OH radical was found to be more efficient in degradation and it followed a pseudo first order kinetics. The generated free radicals remain a strong and non-selective oxidant thus emerging as a powerful oxidant in the degradation of pollutants. Moreover, the acidic media aided the degradation on comparison with the neutral or basic media due to the higher ration of HOCl radicals. Here the confirmation of carbonyl groups in the FTIR analysis (new peak at 1710 cm^{-1}) of degradation products was helpful in proposing a degradation pathway via alcohol to carbonyl. Most of the microorganisms are not evolved to utilize PVA as a carbon source [41, 42].

Radiation-induced degradation implies on using the reactive species generated through the radiolysis of water to oxidize PVA [43]. However, coast remains a major concern with the wide application of this method for PVA degradation. While considering the photocatalytic degradation, the specific conditions required limits its applications [44]. Again, PVA with higher degree of hydrolysis degraded faster than that with lower hydrolysis. This could be due to the higher number of hydroxyl groups which induced higher number of chain scissions which in turn made the degradation easier by reducing the chain size [41]. The degradation mechanism of the polymer can be varied with the applied temperature where, higher degradation can be resulted from a higher temperature. However, the degradation of the thermo-oxidative conditions remain the major mechanism involved. The lifespan predictions involve the accelerated aging studies. This time- temperature relationship can be studied using the data acquired at various aging temperatures. The lifespan estimation according to the international standards (ISO 2578[24]) is aimed at determining the thermal endurance limits of the plastics in general. It basically depends on the tensile properties of the aged material and the 50% loss in the tensile properties is determined as the end point [45].

The morphology and size of the developed fibers can de varied with the type and concentration of the fillers used. The electrospun fibers have found its applications in the food sector considering its ability to stabilize and control the release of the active components. The time dependent changes occurring in the polymer leading to the significant changes in the molecular structure are said as the physical ageing. It can also be affected by other factors such as water which leads to the hydrolytic degradation which can also result in the reduction of molecular weight which results in brittleness. This ageing can be caused due to the molecular motions caused by the non-equilibrium state of the polymer below the glass transition temperature. The polymer degradation can be caused by photochemical or hydrolytic processes as well as the thermal factors. The shear stress imposed on a polymer as a result of the temperature to which the films are subjected can result in the chain scission which in turn can reduce the molecular weight and the intrinsic viscosity of the membranes. The variations in the molecular weight of the samples thus caused can be assessed through the intrinsic viscosity measurements. The polymer degradation can also alter the thermal stability, mechanical and optical properties of the polymer. The intrinsic viscosity of the polymer reduces with increase in the chain scission rate. The intrinsic viscosity profiles of the PLA indicated its ability to withstand a single recycling processes without causing a huge impact to the weight loss over the period. However, the various washing steps introduced can weaken the polymer structure. Thus, the structure and final properties of the PLA can be controlled through the reduction in the molecular weight [46].

The structural changes in the polymer when subjected to the degradation can be confirmed through the FTIR-ATR spectra. The C=O stretching bands in the PLA is indicated through the presence of bands in region 1756 cm^{-1} and can be exploited as an indicator of the degradation. This can be attributed to the generation of new carbonyl-linked species. Moreover, the displacement of the bands corresponding to the carbonyl stretching towards the higher wavelength can be attributed to the various reprocessing steps. However, the analysis has confirmed only minor variations in the PLA subjected to the recycling on comparison to the virgin form indicating the generation of lower number of carbonyl compound generation. The changes in the structure of the polymer while subjected to the recycling can be confirmed using DSC. The heating scans of the native PLA confirms 2 peaks, a glass transition at a temperature of 60°C along with an endothermic peak related to the physical ageing. The recycling does not affect the glass transition temperature; however, an exothermic peak is obtained at the 110°C which corresponds to the cold crystallization temperature. The reduction in the cold crystallization temperature of the recycled materials (about 5°C) can be attributed to the higher mobility of the polymer chains resulted from the reduction in the molecular weight while subjected to the reprocessing. The double endothermal peaks emerged at around 150°C corresponds to the melting of the polymer. This double melting behavior can be attributed to the melt recrystallization mechanism wherein the melting of the less perfect crystals occurs at lower temperatures and that of the more stable ones occur at a higher temperature.

The thermal stability of the material can be analyzed using TGA. It is evident from the studies that the smaller reduction in the molecular weight when subjected to recycling did not affect the thermal stability significantly. However, the generation of shorter polymer chains having presumably low thermal stability can reduce the overall thermal stability of the material. The effect of the reprocessing conditions on the hardness and the Youngs modulus of the PLA is found to be negligible on comparison to the native fibers. Thus, the overall variations caused to the structure along with the thermal and mechanical stability of the PLA is minimal on a moderate degradation process [46]. The higher opacity of the samples subjected to the composting conditions for a week can be attributed to either the variations in the refractive index to the water absorption or the presence of hydrolytic product formation. Colour change and mycelial growth was seen in these films after being subjected to the composting conditioning for 2 weeks or more. The incorporation of nanoparticles (Clay, $CaCO_3$, nano-SiO_2) was found not to hamper with the disintegration of the PLA which occurred in 28 days under composting conditions under the laboratory conditions. However, the higher hydrophobic condition created by the SiO_2 nanoparticle did protect the PLA matrix from the attack of water. Since the degradation of the PLA occurs mainly through the hydrolysis, the hinderance to water caused by these nanoparticles did slow down the process. And a similar trend was observed in the case of other polymers such as PHB as well. However, the end result was a 100% disintegration inspite of the longer time required.

The biodegradation of the polymers can lead to the formation of carbon dioxide, water, inorganic compounds and biomass while consuming oxygen under aerobic conditions. The elemental analysis of carbon can lead to the theoretical quantification of the carbon dioxide formed. A typical degradation pattern of the PLA includes an initial lag phase due to the hydrolysis of ester bonds by the extracellular hydrolytic enzymes leading to the degradation of the generated monomers. This trend can be seen in PLA copolymer films as well. The lag phase however can vary with the incorporation of various nanoparticles. To elaborate further, in case of clay nanoparticles it was 9 days while for the nano-$CaCO_3$ it was 11 and for nano-SiO_2 it was 16-17 days as similar to the blank films. This variation can be due to the higher hydrophilicity of the nanoclay which in turn enhances the water absorption in the polymer matrix thereby enhancing the rate of the hydrolytic degradation. Further, the ecotoxic analysis of the composite did not show any visual signs of phytotoxicity in terms of chlorosis, necrosis, wilting or other deformations.

6.10 Conclusions

Biodegradable polymers are being exploited in all the sectors ranging from food to medicine and industry owing to its beneficial impact on the environment. Eventhough, a large number of polymers are available, their applications are limited by their intrinsic properties. This can be answered through the development of polymer blends containing plasticizers and fillers to obtain the required tailor-made properties. Each of the additional components added to the matrix can have greater impact on the overall functioning of the material. Among the polymers, PVA and PLA find a larger application due to its versatility. The tailoring of the individual polymers has led to their functioning in wider aspects. Moreover, these polymers are biodegradable even in the presence of fillers which has increased its appeal in all sectors.

References

1. Cazón, P., Vázquez, M., Velazquez, G., Cellulose-glycerol-polyvinyl alcohol composite films for food packaging: Evaluation of water adsorption, mechanical properties, light-barrier properties and transparency. *Carbohydr. Polym.*, 195, 432–443, 2018.
2. Almeida, E.V.R., Frollini, E., Castellan, A., Coma, V., Chitosan, sisal cellulose, and biocomposite chitosan/sisal cellulose films prepared from thiourea/NaOH aqueous solution. *Carbohydr. Polym.*, 80, 3, 655–664, 2010. https://doi.org/10.1016/j.carbpol.2009.10.039.
3. Gutiérrez, T.J., Polymers for food applications: News, in: *Polymers For Food Applications*, pp. 1–4, Springer, Cham, 2018.
4. Nagarajan, B., Kamkar, M., Schoen, M.A., Sundararaj, U., Trudel, S., Qureshi, A.J., Mertiny, P., Development and characterization of stable polymer formulations for manufacturing magnetic composites. *J. Manuf. Mater. Process.*, 4, 1, 4, 2020.
5. Bunn, C.W., Crystal structure of polyvinyl alcohol. *Nature*, 161, 4102, 929–930, 1948.
6. Zhang, Y., Rong, W., Fu, Y., Ma, X., Photocatalytic degradation of poly(vinyl alcohol) on Pt/TiO_2 with Fenton reagent. *J. Polym. Environ.*, 19, 4, 966–970, 2011.
7. DeMerlis, C.C. and Schoneker, D.R., Review of the oral toxicity of polyvinyl alcohol (PVA). *Food Chem. Toxicol.*, 41, 3, 319–326, 2003.
8. Liu, M., Guo, B., Du, M., Jia, D., Drying induced aggregation of halloysite nanotubes in polyvinyl alcohol/halloysite nanotubes solution and its effect on properties of composite film. *Appl. Phys. A*, 88, 2, 391–395, 2007.
9. Giroto, J.A., Guardani, R., Teixeira, A.C.S., Nascimento, C.A.O.D., Study on the photo-Fenton degradation of polyvinyl alcohol in aqueous solution. *Chem. Eng. Process. Process Intensif.*, 45, 7, 523–532, 2006.
10. Mathew, S., Jayakumar, A., Kumar, V.P., Mathew, J., Radhakrishnan, E.K., One-step synthesis of eco-friendly boiled rice starch blended polyvinyl alcohol bionanocomposite films decorated with in situ generated silver nanoparticles for food packaging purpose. *Int. J. Biol. Macromol.*, 139, 475–485, 2019a.

11. Mathew, S., Raveendran, A., Mathew, J., Radhakrishnan, E.K., Antibacterial effectiveness of rice water (starch)-capped silver nanoparticles fabricated rapidly in the presence of sunlight. *Photochem. Photobiol.*, 95, 2, 627–634, 2019b.
12. Jayakumar, A., Heera, K.V., Sumi, T.S., Joseph, M., Mathew, S., Praveen, G., Radhakrishnan, E.K., Starch-PVA composite films with zinc-oxide nanoparticles and phytochemicals as intelligent pH sensing wraps for food packaging application. *Int. J. Biol. Macromol.*, 136, 395–403, 2019.
13. Teodorescu, M., Bercea, M., Morariu, S., Biomaterials of PVA and PVP in medical and pharmaceutical applications: Perspectives and challenges. *Biotechnol. Adv.*, 37, 1, 109–131, 2019.
14. Zhou, Y., Lei, L., Yang, B., Li, J., Ren, J., Preparation and characterization of polylactic acid (PLA) carbon nanotube nanocomposites. *Polym. Test.*, 68, 34–38, 2018.
15. Hoidy, W.H., Ahmad, M.B., Al-Mulla, E.A.J., Ibrahim, N.A.B., Preparation and characterization of polylactic acid/polycaprolactone clay nanocomposites. *J. Appl. Sci.*, 10, 2, 97–106, 2010.
16. Saini, P., Arora, M., Kumar, M.R., Poly (lactic acid) blends in biomedical applications. *Adv. Drug Deliv. Rev.*, 107, 47–59, 2016.
17. Vieira, M.G.A., da Silva, M.A., dos Santos, L.O., Beppu, M.M., Natural-based plasticizers and biopolymer films: A review. *Eur. Polym. J.*, 47, 3, 254–263, 2011.
18. Shivakumara, L.R., Demappa, T., Singh, V., Mechanical and barrier properties of hydroxy propyl methyl cellulose edible polymer films with plasticizer combinations. *J. Food Process. Preserv.*, 41, 4, e13020, 2017.
19. Altan, A., Aytac, Z., Uyar, T., Carvacrol loaded electrospun fibrous films from zein and poly (lactic acid) for active food packaging. *Food Hydrocoll.*, 81, 48–59, 2018.
20. Righetti, M.C., Cinelli, P., Mallegni, N., Massa, C.A., Irakli, M., Lazzeri, A., Effect of the addition of natural rice bran oil on the thermal, mechanical, morphological and viscoelastic properties of poly (lactic acid). *Sustainability*, 11, 10, 2783, 2019.
21. Murariu, M. and Dubois, P., PLA composites: From production to properties. *Adv. Drug Deliv. Rev.*, 107, 17–46, 2016.
22. Bognitzki, M., Czado, W., Frese, T., Schaper, A., Hellwig, M., Steinhart, M., Wendorff, J.H., Nanostructured fibers via electrospinning. *Advanced Mater.*, *13*, 1, 70–72, 2001.
23. Koski, A., Yim, K., Shivkumar, S.J.M.L., Effect of molecular weight on fibrous PVA produced by electrospinning. *Mater. Lett.*, *58*, 3–4, 493–497, 2004.
24. Wunner, F.M., Florczak, S., Mieszczanek, P., Bas, O., De-Juan-Pardo, E.M., Hutmacher, D.W., Electrospinning with polymer melts–state of the art and

future perspectives, in: *Comprehensive Biomaterials II*, pp. 217–235, Elsevier, 2017.

25. Wen, H.F., Yang, C., Yu, D.G., Li, X.Y., Zhang, D.F., Electrospun zein nanoribbons for treatment of lead-contained wastewater. *Chem. Eng. J.*, 290, 263–272, 2016.
26. Shi, X., Zhou, W., Ma, D., Ma, Q., Bridges, D., Ma, Y., Hu, A., Electrospinning of nanofibers and their applications for energy devices. *J. Nanomater.*, 6, 7, 129, 2015, 2015.
27. Zupancic, S., Core-shell nanofibers as drug-delivery systems. *Acta Pharm.*, 69, 2, 131–153, 2019.
28. Dror, Y., Salalha, W., Avrahami, R., Zussman, E., Yarin, A.L., Dersch, R., Wendorff, J.H., One-step production of polymeric microtubes by co-electrospinning. *Small*, 3, 6, 1064–1073, 2007.
29. Xiao, C., Zhang, Z., Zhang, J., Lu, Y., Zhang, L., Properties of regenerated cellulose films plasticized with α-monoglycerides. *J. Appl. Polym. Sci.*, 89, 13, 3500–3505, 2003.
30. Tanpichai, S. and Oksman, K., Cross-linked nanocomposite hydrogels based on cellulose nanocrystals and PVA: Mechanical properties and creep recovery. *Compos. Part A Appl. Sci. Manuf.*, 88, 226–233, 2016.
31. Pyrz, W.D. and Buttrey, D.J., Particle size determination using TEM: A discussion of image acquisition and analysis for the novice microscopist. *Langmuir*, 24, 20, 11350–11360, 2008.
32. Guo, J., Ge, L., Li, X., Mu, C., Li, D., Periodate oxidation of xanthan gum and its crosslinking effects on gelatin-based edible films. *Food Hydrocoll.*, 39, 243–250, 2014.
33. Imran, M., Revol-Junelles, A.M., René, N., Jamshidian, M., Akhtar, M.J., Arab-Tehrany, E., Desobry, S., Microstructure and physico-chemical evaluation of nano-emulsion-based antimicrobial peptides embedded in bioactive packaging films. *Food Hydrocoll.*, 29, 2, 407–419, 2012.
34. Burgos, N., Armentano, I., Fortunati, E., Dominici, F., Luzi, F., Fiori, S., Kenny, J.M., Functional properties of plasticized bio-based poly (lactic acid) _ poly (hydroxybutyrate) (PLA_PHB) films for active food packaging. *Food Bioprocess Technol.*, 10, 4, 770–780, 2017.
35. Goh, K., Heising, J.K., Yuan, Y., Karahan, H.E., Wei, L., Zhai, S., Fane, A.G., Sandwich-architectured poly (lactic acid)–graphene composite food packaging films. *ACS Appl. Mater. Interfaces*, 8, 15, 9994–10004, 2016.
36. Yang, C., Zhu, B., Wang, J., Qin, Y., Structural changes and nano-TiO_2 migration of poly (lactic acid)-based food packaging film contacting with ethanol as food simulant. *Int. J. Biol. Macromol.*, 139, 85–93, 2019.

37. Kobayashi, M. and Hyu, H.S., Development and evaluation of polyvinyl alcohol-hydrogels as an artificial atrticular cartilage for orthopedic implants. *Materials*, 3, 4, 2753–2771, 2010.
38. Ma, S., Wang, S., Li, Q., Leng, Y., Wang, L., Hu, G.H., A novel method for preparing poly(vinyl alcohol) hydrogels: Preparation, characterization, and application. *Ind. Eng. Chem. Res.*, 56, 28, 7971–7976, 2017.
39. Scaffaro, R. and Lopresti, F., Processing, structure, property relationships and release kinetics of electrospun PLA/Carvacrol membranes. *Eur. Polym. J.*, 100, 165–171, 2018.
40. Kawai, F. and Hu, X., Biochemistry of microbial polyvinyl alcohol degradation. *Appl. Microbiol. Biotechnol.*, 84, 2, 227, 2009.
41. Ye, B., Li, Y., Chen, Z., Wu, Q.Y., Wang, W.L., Wang, T., Hu, H.Y., Degradation of polyvinyl alcohol (PVA) by UV/chlorine oxidation: Radical roles, influencing factors, and degradation pathway. *Water Res.*, 124, 381–387, 2017.
42. Chiellini, E., Corti, A., D'Antone, S., Solaro, R., Biodegradation of poly(vinyl alcohol) based materials. *Prog. Polym. Sci.*, 28, 6, 963–1014, 2003.
43. Sun, W., Tian, J., Chen, L., He, S., Wang, J., Improvement of biodegradability of PVA-containing wastewater by ionizing radiation pretreatment. *Environ. Sci. Pollut. Res.*, 19, 8, 3178–3184, 2012.
44. Lin, C.C., Lee, L.T., Hsu, L.J., Degradation of polyvinyl alcohol in aqueous solutions using UV-365 nm/$S_2O_8^{2-}$-process. *Int. J. Environ. Sci. Technol.*, 11, 3, 831–838, 2014.
45. Anakabe, J., Santamaria-Echart, A., Eceiza, A., Arbelaiz, A., Zaldua Huici, A.M., Evolution of the mechanical properties and estimation of the useful lifespan of poly (lactic acid) based compounds. *Polym. Int.*, 67, 6, 761–769, 2018.
46. Beltrán, F.R., Lorenzo, V., Acosta, J., de la Orden, M.U., Urreaga, J.M., Effect of simulated mechanical recycling processes on the structure and properties of poly (lactic acid). *J. Environ. Manage.*, 216, 25–31, 2018.

7

Biomedical Applications of Polyvinyl Alcohol-Based Bionanocomposites

Bruno Leandro Pereira[1,2], Viviane Seba Sampaio[1,3], Gabriel Goetten de Lima[1,4], Carlos Maurício Lepienski[4,5], Mozart Marins[3], Bor Shin Chee[1] and Michael J. D. Nugent[1]*

[1]*Materials Research Institute MRI, Athlone Institute of Technology, Athlone, Ireland*
[2]*Department of Mechanical Engineering, Pontifícia Universidade Católica do Paraná, Curitiba, PR, Brazil*
[3]*Biotechnology Unit, University of Ribeirão Preto, Ribeirão Preto, SP, Brazil*
[4]*Federal University of Paraná, Postgraduate Program in Materials Engineering and Science - PIPE, Curitiba, PR, Brazil*
[5]*Federal University of Technology - Paraná, Department of Mechanical Engineering, Curitiba, PR, Brazil*

Abstract
Polyvinyl alcohol (PVA) is a synthetic biodegradable polymer that can mimic natural polymers and has excellent characteristics of biocompatibility in human tissues. As a polymeric matrix, PVA properties can also be enhanced or changed by the introduction of nanofillers. Combining PVA and a variety of nanofillers, it is possible to create materials with unique characteristics with many applications in the biomedical field such as drug delivery, wound healing, tissue engineering and medical devices. This chapter will focus on a review of literature about the research of nanocomposites using PVA as a matrix and the properties possible by the introduction of a range of different nanofillers for biomedical usage.

***Keywords*:** Polyvinyl alcohol, biocompatible nanocomposite, nanofillers

**Corresponding author*: Michael.Nugent@tus.ie

Visakh P. M. and Olga B. Nazarenko (eds.) Polyvinyl Alcohol-Based Biocomposites and Bionanocomposites, (179–204) © 2023 Scrivener Publishing LLC

7.1 Introduction

Polymers have a large range of applications due to the various properties that can be obtained within these materials such as corrosion resistance, flexibility, softness/stiffness, and density (low mass concentration) [1]. However, polymers have known disadvantages, such as poor thermal and mechanical properties that limit their usage [1, 2]; methods to overcome these limitations includes the incorporation of nanoparticles into the polymeric matrices to intentionally form nanocomposites [3]. Nanocomposites can be separated into two parts nano and composite. The definition of nano refers to the material size scale (10^{-9} m), meaning that the material has, at least, a dimension less than 100 nm [4]. While the word composite is frequently defined as multiphasic materials that display a substantial part on the characteristics of component phases to specifically aim for better properties in certain applications [4]. Polymer nanocomposites are composed of nano-sized phase (or phases) dispersed uniformly in a polymeric matrix [5, 6], and they can exhibit formation of layers, fibers, clusters, or particles obtained from organic or inorganic sources [7]. Nonetheless, one of the great advantages of nanoparticles are the large surface area exposure; therefore, only a relatively small volume of filler is needed to enhance the material properties [8]. Dispersion of nanofillers is normally carried out by ultrasonic bath or mechanical stirring, and the greatest challenge is to avoid agglomerations. One approach is to use techniques to decrease its viscosity such as using melted polymers [9].

Biocompatible nanocomposites can be formed using as matrix biocompatible polymers, such as polyvinyl alcohol (PVA) [10]. PVA is a synthetic biodegradable polymer that can mimic natural polymers [3], due to its characteristics of biocompatibility in human tissues, high hydrophilicity, non-toxicity, corrosion resistance, mechanical flexibility, excellent optical transmission, water solubility, and non-mutagenic behavior [11–15]. Additionally, PVA properties can also be enhanced or changed by the introduction of nanofillers in this polymer. The employed technique is also an important parameter to improve target properties and some of these techniques includes electrospinning [16, 17], solvent casting [10, 18], metal ion coordination [19], freeze-casting followed by freeze-drying [20], lyophilisation [21, 22], thermal cycles [23], ultra-sonication process [24], and salt leaching [25].

Nanocomposites using PVA have many applications in the biomedical field, including drug delivery systems [26–28], wound healing, tissue engineering [29], and regenerative medicine. When compared with pure PVA, PVA nanocomposites may improve protection against bacteria, thermal stability [30, 31], mechanical properties [32], and bioactivity [29]. This chapter will focus on design and challenges of nanocomposites using PVA as a matrix in biomedical applications.

7.2 Application in Drug Delivery Systems

The drug delivery system is a method or process that allows targeting with effective delivery of a particular drug to achieve a therapeutic effect; nonetheless, these systems possess shortcomings that requires attention. Many drugs lose or have their effect diminished due to degradation, and this can occur even before they reach the correct destination/path [33]. Therefore, it is necessary to develop a methodology that delivers these drugs while preserving their pharmacological properties during the journey inside the body, as well as to avoid the release in areas that are not within their target destination [33]. Another important factor is the administration of drugs level within the bloodstream and commonly, as soon as the drug is administered, the initial concentration of the drug in the body is high, and this accumulation decreases quickly over time–burst release. Consequently, after a certain period, this drug is no longer effective and is one of the main challenges in this field [34, 35].

Controlled delivery systems allow for the delivery of medicines in a prolonged manner, avoiding undesirable oscillations in drug concentration, and thus maintaining the optimal therapeutic drug concentration in systemic circulation for an extended period. In addition to controlled release, these systems may be modified for a target release of the drug. This target release allows for the delivery of therapeutic agents directly to an organ or tissue, increasing drug efficacy, as well as reducing its toxicity [35].

The research field of controlled target drug delivery is particularly attractive, and researchers have provided solutions for this specific application. Therefore, the use of materials that respond to stimuli produced specifically at the site where the drug is planned to be delivered has been extensively studied. The reason behind using these materials is that no release of the drug will occur unless it overcome the threshold of pre-established stimuli. These stimuli may be intrinsic, such as pH, different types of enzymes, and proteins, among others [36, 37] and extrinsic stimuli may also be applied, such as temperature, ultrasound, magnetism, among others [38]. To avoid harmful side effects, controlled delivery systems are very useful. Using targeted drug delivery these side effects will not occur or at least be minimized [39].

Controlled and Targeted Drug Delivery Systems will likely replace the currently used systems, due to their many advantages, some of which are highlighted below:

- enhancement in therapeutic effectiveness,
- improvement of drug solubility,
- improvement of pharmacokinetic profile and bioavailability,
- protection against degradation in the gastrointestinal tract and/or plasma,
- increase in plasma concentrations or in the target tissue of short half-life drugs,
- reduction in the amount of drug dosage,
- reduction in administration of the drug,
- safer usage of highly potent drugs,
- site-specific drug release, decreasing drug toxicity in non-target tissues or organs,
- significant reduction on local and systemic toxicity,
- reduction on the treatment cost.

The development of efficient controlled release systems essentially depends on the selection of a carrier system capable of regulating the release of the drug and bypassing the limiting physicochemical properties of the transported drug [35].

Polymer research on the development of drug delivery systems is due to the fact that the polymer acts as a barrier to the immediate release of the drug [40]. However, it is worth mentioning that special attention to the solubility and diffusion of the drug with the polymer is required because interactions between drug and polymer may affect the rate of release as well as interactions between the drug and other components used. These interactions may affect solubility in water, which in turn, affects the drug release rate [41, 42]. Drug release by the polymer depends on several factors: drug solubility, dispersibility, release mechanism, polymer degradation and the combination of processes of breakdown and propagation [43].

The main release methods for active ingredients, present in the polymer, to be released consists of three ways: diffusion, degradation, and swelling. Diffusion occurs when the drug incorporated into the polymer reaches the medium through pores in the polymer matrix, as this movement continues to occur, the release rate decreases due to the reduction in the saturation of the drug within the polymer matrix [44]. In the degradation process, the polymer is degraded within the body via hydrolyzation of the long polymer chains to progressively transform into smaller compounds. Usually, there are three steps involving into this release: drug release from the surface, release from early degradation of the polymer matrix, and release of remaining drug during complete degradation of the polymer matrix [45]. Lastly, swelling refers to the process in which the polymers, when placed in the body, have the capacity to absorb water, as well as body fluids and swell. In this way, the pores of the polymer matrix are increased allowing the drug to be released to the external environment [46]. Materials that absorb large quantities of water and swell to release a specific drug are known as hydrogels.

Applications of nanocomposite polymer concepts in drug delivery systems are worth investigating at aiming special drug properties by the creation of smart drugs. Depending on the nanofiller inside the polymeric matrix, it is possible to modify the rate of drug release; as an example, the electrostatic interactions between the drug and nanofiller may cause a decrease in release rate when the drug and the nanofiller have opposite charges [47]. Among other notable interactions is the one that happens between the nanofiller/drug and the biological tissue, and to minimise the drug side effect is desired that the drug ideally acts only in the target tissue. This can be achieved using nanofiller sensitive to external stimuli, such as a magnetic field to deliver the drug or using nanofillers with affinity with the unhealthy target tissue [47, 48]. Another alternative is to alter the polymeric matrix properties by the nanofiller such as swelling and drug release rate [49]. Among the nanofillers, carbon nanotubes (CNT) [50, 51], graphene oxide (GO) [52], copper oxide [49] and superparamagnetic iron oxide nanoparticles (SPIONs) [53, 54] present interesting application in drug delivery as nanofillers.

Carbon nanotubes CNT are non-immunogenic, biocompatible, presenting a higher capacity of accumulation in the tumor when compared with healthy tissue, and are capable of carrying high drug dosages [50, 51]. By the functionalization process of CNT, this nanofiller is able to deliver hydrophilic and lipophilic drugs, crossing barriers in cellular level, such as plasma membrane [55]. Thus, CNTs present great potential in the use of drug delivery systems. In a nanocomposite made with polyvinyl alcohol (PVA) and polyvinylpyrrolidone (PVP) an increase in the drug release amount caused by CNT addiction into the nanocomposite matrix was observed [56].

Graphene oxide (GO) and its associated compounds exhibit a great potential to drug carrier in cancer treatments [57]. In a PVA nanocomposite filled with GO, a pH sensitivity was detected in drug release process [52]. Superparamagnetic iron oxide nanoparticles (SPIONs) are biocompatible, non-toxic, and biodegradable [53, 54]. The sensitivity of magnetic nanoparticles to the external magnetic field is useful to reach specific biological tissue targets conducting the drugs using magnetic forces [53]. If the drug is linked to SPION, it can decrease the side effect by keeping the drug concentration only in the target tissue and decreasing the drug concentration in healthy tissues. Another aspect is the superparamagnetic phenomenon, which can avoid aggregation of nanoparticles; when the external magnetic field is removed, the magnetic force between the nanoparticles is negligible [54]. Another use of the SPION is the production of thermosensitive nanocomposites. It was demonstrated that PVA has an important role in a thermosensitive triblock-copolymer matrix to avoid drug leakage and provides more stability. This nanocomposite was prepared by mini-emulsion technique with drug and iron oxide nanoparticles inside of the copolymer. The formed nanocomposite showed benefits as nanocarriers thermosensitive to an external magnetic field for the dealing of acute diseases [58]. The cisplatin is a powerful drug to fight against cancers; however, it has undesirable side effects. To minimize these problems, it was proposed a nanocomposite with a core composed by PAA and cisplatin covered by a shell composed by PVA/iron oxide. The magnetic targeting results in nine times more accumulation in tumour tissue than the pure drug, showing an excellent rate of tumour decreasing after 24 days with a negligible side effect in comparing with a control group [59].

7.3 Applications in Wound Healing

A wound can be defined by a rupture in the continuity of skin or mucosa, which can be chronic or acute (8–12 weeks to wound to heal) depending on the category of healing process [60]. The wound can result from physical, chemical, thermal damage or pathological process. The healing process is a complex, ordered and dynamic course starting with coagulation, inflammation, proliferation period (formation of new tissues—extracellular matrix, epithelisation, angiogenesis, and contraction), and the remodeling phase [60, 61]; due to the large spectra of different kind of wounds, different approaches must be considered.

The wound dressing is a treatment to help the healing process [62]. Currently similar techniques (covering the wounds) are being employed, and many types of research aim at creating the best conditions to optimise and accelerate the healing process. The ideal wound dressing, among other properties, has to be non-toxic, non-adherent, antibiotic, gas permeable, mechanically suitable, and additionally supply humidity and absorb exudates [62, 63].

Hydrogels present a good behavior to be applied in wound healing area due to hydrophilic property, nontoxic, biodegradable flexibility, swelling by responding environment stimuli, the capability to incorporate exudates and act as a barrier to infections during the healing process, allowing drug delivery through the membrane [30, 64, 65]. These characteristics provide an excellent application of PVA hydrogels in the wound dressing field. However, hydrogels made solely of PVA do not have suitable mechanical properties, which limit the applications. Creating PVA nanocomposites is an excellent alternative to address this issue, and desired nanocomposites properties are directly dependent on the nanofiller and the production method. Table 7.1 shows the introduced nanofiller in the PVA matrices with the respective properties or/and improvements.

Skin is a natural wall that protects the body against bacterial invasion and infections and if the skin tissue is injured, microorganism can develop. Certainly, one of the concerns about would dressing production is the antibacterial property [79]. Silver is utilised to keep the environment free from bacterial infection in a large range of different species [80]; however, silver compounds can be harmful at high levels or long-time exposure [81]. PVA Silver nanocomposites can keep the metal release under control [82]. Using silver nanoparticles with carbon nanotubes as PVA nanofillers, the antibacterial behavior can be extending up to 72 hours [82]. The optical properties of PVA/silver nanoparticles can be used to create a transparent wound dressing, which maintain the wound visible to medical diagnose with the injury free from bacteria for longer periods [66].

Table 7.1 The improvements and/or properties occasioned by different nanofillers in PVA nanocomposites in wound dressing applications.

Nanofiller	Properties/improvements
Silver [38]	Anti-bacterial
Carbon nanotube + Silver [39]	Extended anti-bacterial life-time
Quaternized PVA synthetized with silver nanoparticles [66]	Extended anti-bacterial life-time
Propolis [67]	Anti-bacterial
Curcumin [47]	Anti-bacterial
Curcumin + silver [68]	Anti-bacterial
Calcium alginate [69]	Anti-bacterial
Na alginate + bioglass [70]	Mechanical properties, degradation, bioactivity.
Nanoclay [71–73]	Less flammability, tensile modulus, hardness, and elongation at break.
Na-montmorillonite [71, 74]	Tensile modulus, elongation, bioactivity, swelling performance, antibacterial activity.
Bacterial cellulose [75, 76]	Elastic modulus, tensile strength, non-cellular adhesion.
Zinc monoxide ZnO [77, 78]	Mechanical properties, antibacterial activity, and cellular viability.

One other additive of interest is propolis. Propolis is a product obtained from the production of honeybees used in structural natural propose in their beehives and preventing local bacterial activity. The substance is complex, having more than 300 compounds identified and the composition, among other factors, depending on the season, region, and plant source [83]. This substance has interesting therapeutic properties, such as anti-viral, anti-fungal, anti-bacterial, anti-cancer, anti-inflammatory, anti-protozoal, anti-hepatotoxic, anti-mutagenic, and anti-oxidant [83–87]. Propolis extract blended PVA prepared by electrospun technique presented also an efficient bactericide behavior against gram-positive *S. aureus* [67]. The structure produced is porous due to the electrospinning process and mimics the extracellular matrix [70]. Figure 7.1 exhibits a porous structure formed with PVA/propolis produced by the electrospinning process.

Curcumin is a natural compound obtained from turmeric with anti-inflammatory, anti-cancer (toxic to different cancer cells), and anti-oxidant properties. However, curcumin is not water-soluble, which represent a challenge to blend with PVA. Kamel and co-workers used a method to produce a PVA/Chitosan/Curcumin nanocomposite. The nanocomposite showed nontoxic and excellent bactericide effect [88]. Another study using Curcumin encapsulated Chitosan-PVA silver nanocomposite shows great potential to wound therapy by the better anti-microbial activity against *E. coli* bacteria when compared with chitosan-PVA silver nanoparticles and curcumin only [68].

Alginate is a polymer used in pharmaceutical drugs, wound healing, and when mixed with PVA present non-irritation effect on the eye and skin tissue [70]. Calcium alginate PVA nanocomposites prepared by electrospinning presented in rat wound, healing capacity, and showed a direct

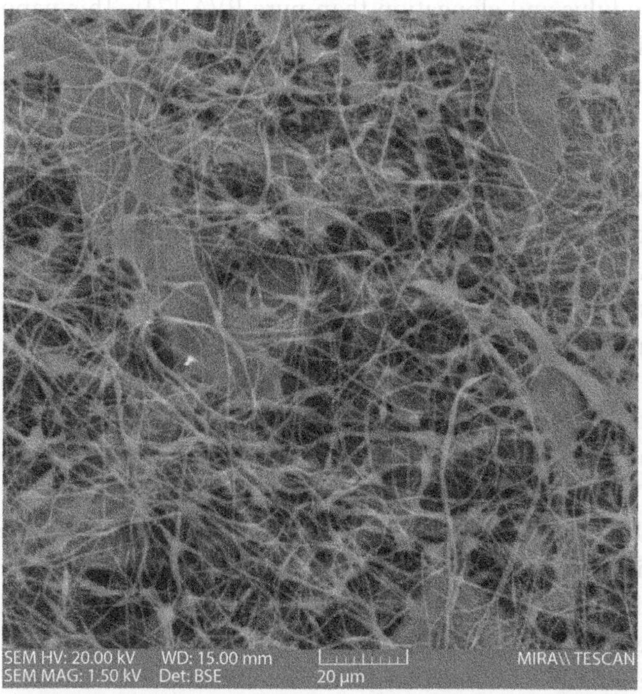

Figure 7.1 Scanning electron micrograph porous structure formed with PVA/propolis produced by electrospinning process.

relationship between the calcium alginate content and the antibacterial activity against to Staphylococcus aureus [69]. A porous nanocomposite made combining Na alginate, bioglass, and PVA prepared by electrospinning technique displayed improved mechanical properties and bioactivity when compared with PVA nanofibers [70].

Mechanical properties of PVA nanocomposites can be enhanced filling with nanoclay [71, 72], nanoclay into PVA matrices has demonstrated improvements, such as less flammability [73]. The tensile modulus, hardness, and elongation at break presented enhancements, increasing the nanoclay percentage in the PVA matrix hydrogel nanocomposites prepared by freeze-thaw method [72]. In addition, 3mm of PVA-nanoclay layer can keep the environment sterile for 1 week. Comparatively, even more than 60 gauze sheets have this level of protection against bacteria [72]. A natural Na-montmorillonite (hydrophilic nanoclay) mixed with PVA showed an inverse relation between of the nanoclay percentage amount and dehydration rates, and a direct relation with the temperature [89]. In addition, low amounts of Na-rich montmorillonite in PVA matrix exhibited better tensile modulus and elongation than pure PVA [74]. This nanofiller also enhanced mechanical properties, swelling performance, biocompatibility, antibacterial activity [71], and did not affect the A-431 cell line (human skin cell) viability blended with PVA/Chitosan nanocomposites [90].

Bacterial cellulose is a renewable natural polymer produced by bacterial activity with stimulating properties, such as biocompatibility and excellent mechanical properties [75, 91]. For example, the bacteria *Gluconacetobacter xylinus* have an excellent polymerization degree forming long microfibrils, which can be "break" by a chemical and mechanical process to form nanocrystals and be incorporated in PVA matrices. Formed nanocomposites using the aforementioned bacterial cellulose nanocrystals can improve the elastic modulus and tensile strength of PVA nanocomposites [75]. Another interesting way to produce nanocomposites is to produce bacterial cellulose in PVA presence by fermentation procedure followed by a crosslink process. The PVA/bacterial cellulose showed suitable results when a cellular adhesion needs to be avoided [76].

The addition of zinc monoxide (ZnO) nanoparticles in PVA followed by freeze-thaw method displayed an enhancement in the mechanical properties, antibacterial activity, and cell viability with the ZnO increase (5 wt.%, 10 wt.%, and 15 wt.%—based on PVA weight). On the other hand, the swelling behavior decreased [77]. However, the filler addiction have a limit, from above this point the mechanical properties decreases. A study involving nano ZnO immobilized in alginate-PVA produced by solvent casting evaporation followed the proportions from 5% to 30% w/v. The optimized value for best mechanical properties was 20% w/v and the bactericide properties against *Staphylococcus aureus* and *Escherichia coli* enhanced with ZnO increment [78].

7.4 Applications in Tissue Engineering

The healing process normally occur in its five ways exhibited in Figure 7.2. The tissue engineering is one of these methods, and nanocomposites for tissue engineering will be the subject of this section [92]. Tissue engineering is an interdisciplinary area that links science/engineering with the biological process at achieving the objectives of reconstruction, maintenance or enhancement of tissue function through biological substitutes [93]. The biological substitutes can be made of scaffolds and living cells [92].

Tissue engineering is majority focused in 3D scaffold development, and these structures are normally treated to form tissue by delivery of cells, growth factor or by the formation of a bioreactor [93]. For the production of the scaffold, the needed requirements are biocompatibility, biodegradability, suitable mechanical properties, porosity, biointegration, and clinical, industrial, and commercial viability [92, 93]. As mentioned previously, PVA is a biocompatible polymer being an interesting material to build scaffolds and to improve the suitable properties for tissue engineering applications; therefore, to build these nanocomposite scaffolds,

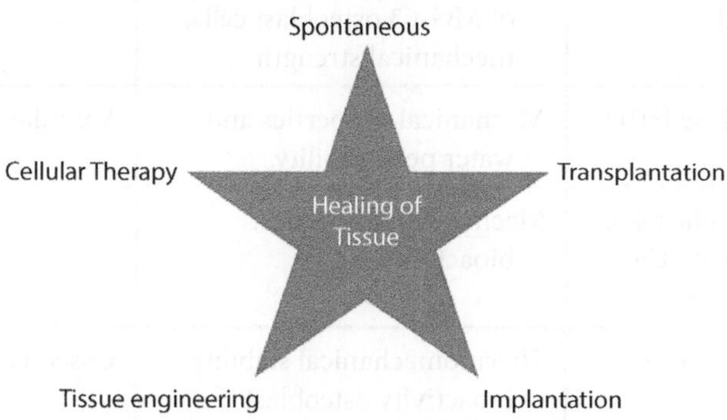

Figure 7.2 The ways of healing tissue process normally occurs [92].

the nanofiller and the production technique have a crucial role to reach the adequate mechanical property, controlled degradation rate, and pore structure. Table 7.2 exhibited some examples of nanofillers relating to the improvements and tissue applications.

The conductivity of polymers in tissue engineering has demonstrated great potential in muscle skeletal application due to a positive response by myoblast (precursor cell of muscular fiber) [99].

Graphene oxide, reduced graphene oxide, and graphene are matrix fillers with a large application in nanocomposites to improve mechanical properties, such as elastic modulus, compressive and tensile strength, conductivity and also present good cytocompatibility with cell bone. Due to hydrogen bonds with the polyvinyl alcohol, graphene oxide presents better behavior when with PVA in heated aqueous solutions than graphene; however, frequently graphene nanocomposites show better mechanical properties

Table 7.2 The improvements and tissue application occasioned by different nanofillers in PVA nanocomposites.

Nanofiller	Improvements	Tissue application
Graphene-based materials [18, 19, 21, 24, 94–96]	Elastic modulus, tensile strength, ductility, electrical conductivity, *angiogenesis*, and *arteriogenesis*.	Osseous, vascular, neural, cartilage
Carbon nanotube [97–99]	Structural properties, porosity, morphology, cellular proliferation.	Neural, osseous
Hydroxyapatite [17, 100]	adhesion and proliferation of MG-63 osteoblast cells, mechanical strength.	Osseous
Nanocellulose [101]	Mechanical properties and water permeability.	Vascular, cartilage
Zirconium phosphate (doped with Ca, Mg, Ti) [16]	Mechanical properties, bioactivity.	
Titanium dioxide [102]	Thermomechanical stability, bioactivity osteoblast adhesion.	Osseous

[24, 94, 95, 103]. Therefore, nanocomposites of PVA and graphene oxide have great characteristics for scaffolds to be used in osseous, neural and muscular (owing to enhance charge carries conductivity) tissue, improvements (or reduction) of new blood vessel formation, depending partially on the fillers amount [21, 104–106].

Carbon nanotubes (CNT) are sheets of graphite wrapped in a tube form with one dimension on the nanoscale (tube diameter), being possible configurations of wall tubes with one single layer or multiple layers [107]. As polyvinyl alcohol filler, CNT in small amounts can cause excellent improvements on mechanical, biological, and electrical properties in PVA nanocomposites [22, 99]. The biological properties include a decrease of bone tissue healing time and biocompatibility are attributed to a CNT large surface area. When the CNT are in agglomerated states these fillers presented cellular human cytotoxicity. In addition, it seems to have a limit amount (1% wt.) on the percentage of CNT to which the mechanical and cellular (MG 63 osteoblast cells) performance seems to be optimised [22]. With the previously cited characteristics, it is possible to conclude that PVA nanocomposites based on carbon nanotubes have a large potential application in engineering tissue, such as in osseous, neural, and muscular tissue.

Osseous tissue is a natural composite composed mainly by carbonated hydroxyapatite ($Ca_5 (PO_4, CO_3)_3 (OH)$) – HAp-dispersed in a polymer (collagen) [108]. To mimic some characteristics of natural bone, hydroxyapatite nanoparticles (a very similar calcium phosphate to bone mineral portion) can be used as filler in polymeric matrices for tissue engineering. Nanosized hydroxyapatite has a hydrophilic behavior [109], which is a good characteristic to mix with PVA to produce nanocomposites. A relative small amount of HAp can improve mechanical strength [100]. Studies involving PVA with nanosized HAp, and polycaprolactone (PCL) showed good adhesion and proliferation of MG-63 osteoblast cells. Despite exceptional biocompatibility in bone tissue, Hap also presents excellent biocompatibility in soft tissues, such as gum, skin, corneal, and muscle [110, 111]. The combined properties of Hap and PVA present a potentially large field of application in human tissues.

Bacterial nanocellulose can cause enhancements when forming nanocomposites with PVA. These improved characteristics are mechanical properties, water permeability, suggesting good application on blood vessels scaffolds [101] and articular cartilages [23]. An interesting relation was found between filler of cellulose and hydroxyapatite, while cellulose nanofibers decelerate the degradation process, the nanoparticles of hydroxyapatite cause the inverse effect [20].

Fillers of zirconium phosphate (doped with Ca, Mg, and Ti) nanoparticles promoted, also, improvements in mechanical properties, controlled degradation, and bioactivity superior to pure PVA [29]. Titanium dioxide TiO_2 (3%wt.) nanosized, by hydrogen bonds interaction with PVA when mixed with PVA, presented enhancements on mechanical, thermostability, and osteoblast cell adhesion [102].

7.5 Applications in Regenerative Medicine

Although regenerative medicine deals with methods to develop regrowing, repair or replace the damaged or diseased cells, organs or tissues; it includes the use of therapeutic stem cells, tissue engineering and production of artificial organs. Therefore, research on regenerative medicine using nanocomposites is highly attractive, and it can show an improvement based on the conventional treatments [112]. In addition, there are many reports on the use of PVA matrix polymer with several nanocomposites for this specific field, such as nano-chitosan [113], -hydroxyapatite [114], and halloysite nanotubes [115], suggesting that these PVA-based nanocomposites can further improve cell regeneration with higher cellular activity for scaffolds with nanofillers. In addition, the different cellular structure and layers, such as cartilage, require materials to have a multiplayer structure to mimic the properties of these various zones. Reports on multilayered hydrogels with nanocomposites for mimicking tissues with various layers, such as cartilage, exhibit an improvement in cell proliferation and a more homogeneous matrix formation [116].

Design for cellular-based therapies is extremely challenging as cell behaviors are reliant on a wide range of scaffold-dependent environmental factors, including cell adhesion, chemical composition, cell receptor stimulation among many others [117]. These factors are the major contributors to an effective cell activity *in vitro* tissue engineering due to the fact that when cells are proliferating onto this polymeric matrix, they produce the extracellular matrix that gradually replaces the scaffold material [118]. Nonetheless, there is an emerging field of cell membrane engineering with few examples of successful modifications on cell membrane have been used to drive *in situ* scaffold formation [119].

The variations in the scaffold properties, such as size and composition, may affect the cell behavior, signal transduction and the polymeric matrix integration with the damaged tissue [120, 121]. As an example, the stiffness and pore size of the scaffold may affect the infiltration, proliferation, and differentiation of the targeted cell [122, 123]. Therefore, there have been currently methods to produce 3D scaffolds, made of the same material, using different technologies to understand the effect of each parameter on the therapeutic outcome [124].

Methods for implantation of these materials can be performed via injectable hydrogels [125] and cell sheets [126] or tissue patches depending if the region cannot be treated with non-invasive methods [127]. Nonetheless, the application of nanocomposites in regenerative medicine has been expanding to the idea of incorporating the sensor capabilities of hydrogels to produce scaffolds that, driven by multiple external stimuli, can act as a small-scale bioinspired soft-robots [128].

7.6 Conclusions and Future Perspectives

Polyvinyl alcohol (PVA) is an excellent material with substantial number of applications in various fields, including human body due an excellent biocompatibility. However, the use of pure PVA presents limitations that can be enhanced by specifics treatments and addition of other compounds. The addition of other compounds can create PVA composites, which may change and enhance its raw properties from the filler nature and size. Nanosized fillers presents advantage by the high interaction with the matrix due to a large surface area exposure and small amounts of nanofillers which may significantly change the nanocomposite properties. This chapter presented detailed nanocomposites with different nanofillers for the application of drug delivery, wound healing, tissue engineering, and regenerative medicine demonstrating the improvements related to the nanofiller. Despite the nature and size of filler to interfere in the properties, other variables, such as filler amount, may change the nanocomposite characteristics. Having a considerable number of variables, new synthetized nanofillers, and different production method mean it is possible to synthesize novel PVA nanocomposites. The future of PVA nanocomposites is linked with the concepts of maximum efficiency, smart materials, and bioadaptability with optimized healing time, multifunctional, painless application, mimicking physical, chemical, and biological environment to successfully return lost biological function.

References

1. Goyal, A., Sharma, A., Saini, I., Chandak, N., Sharma, P., Tailoring of optical and electrical properties of PMMA by incorporation of Ag nanoparticles. *Bull. Mater. Sci.*, 40, 615–621, 2017.
2. Meyers, M. and Chawla, K., Edition: Second. *Mechanical Behavior of Materials*, Cambridge University, 2009.
3. Liu, M., Guo, B., Du, M., Jia, D., Drying induced aggregation of halloysite nanotubes in polyvinyl alcohol/halloysite nanotubes solution and its effect on properties of composite film. *Appl. Phys. A*, 88, 391–395, 2007.
4. Callister, W.D., *Materials Science and Engineering: An Introduction*, John Wiley & Sons, New York, 2007.
5. Kyriacos, D., Chapter 17-Polycarbonates, in: *Brydson's Plastics Materials*, Eighth Ed, M. Gilbert (Ed.), pp. 457–485, Butterworth-Heinemann, Oxford, 2017, https://doi.org/10.1016/B978-0-323-35824-8.00017-7.
6. Kuila, T., Tripathy, T., Lee, J.H., Polyolefin-based polymer nanocomposites, in: *Advances in Polymer Nanocomposites*, pp. 181–215, 2012.
7. Huang, T.-C., Yeh, J.-M., Lai, C.-Y., Polymer nanocomposite coatings, in: *Advances in Polymer Nanocomposites*, pp. 605–638, 2012.
8. Aslam, M., Kalyar, M.A., Raza, Z.A., Polyvinyl alcohol: A review of research status and use of polyvinyl alcohol based nanocomposites. *Polym. Eng. Sci.*, 12, 2119–2132, 2018.
9. Reynaud, E., Gauthier, C., Perez, J., Nanophases in polymers. *Rev. Met. Paris*, 96, 169–176, 1999.
10. Rescignano, N., Fortunati, E., Montesano, S., Emiliani, C., Kenny, J.M., Martino, S., Armentano, I., PVA bio-nanocomposites: A new take-off using cellulose nanocrystals and PLGA nanoparticles. *Carbohydr. Polym.*, 99, 47–58, 2014.
11. Paradossi, G., Cavalieri, F., Chiessi, E., Spagnoli, C., Cowman, M.K., Poly(vinyl alcohol) as versatile biomaterial for potential biomedical applications. *J. Mater. Sci. Mater. Med.*, 14, 687–691, 2003.
12. Ramezanzadeh, B., Vakili, H., Amini, R., The effects of addition of poly(vinyl) alcohol (PVA) as a green corrosion inhibitor to the phosphate conversion

coating on the anticorrosion and adhesion properties of the epoxy coating on the steel substrate. *Appl. Surf. Sci.*, 327, 174–181, 2015.

13. Hikku, G.S., Jeyasubramanian, K., Venugopal, A., Ghosh, R., Corrosion resistance behaviour of graphene/polyvinyl alcohol nanocomposite coating for aluminium-2219 alloy. *J. Alloys Compd.*, 716, 259–269, 2017.
14. DeMerlis, C. and Schoneker, D., Review of the oral toxicity of polyvinyl alcohol (PVA). *Food Chem. Toxicol.*, 41, 319–326, 2003.
15. Caló, E. and Khutoryanskiy, V.V., Biomedical applications of hydrogels: A review of patents and commercial products. *Eur. Polym. J.*, 65, 252–267, 2015.
16. Chee, B.S., De Lima, G.G., Devine, D., Nugent, M.J.D., Electrospun hydrogels composites for bone tissue engineering, in: *Applications of Nanocomposite Materials in Orthopedics*, 813740, 2018.
17. Maheshwari, S.U., Samuel, V.K., Nagiah, N., Fabrication and evaluation of (PVA/HAp/PCL) bilayer composites as potential scaffolds for bone tissue regeneration application. *Ceram. Int.*, 40, 8469–8477, 2014.
18. Bao, C., Guo, Y., Song, L., Hu, Y., Poly(vinyl alcohol) nanocomposites based on graphene and graphite oxide: A comparative investigation of property and mechanism. *J. Mater. Chem.*, 21, 13942–13950, 2011.
19. Lin, C., Liu, Y., Xie, X., GO/PVA nanocomposites with significantly enhanced mechanical properties through metal ion coordination. *Chin. Chem. Lett.*, 30 1100–1104, 2019.
20. Mirab, F., Eslamian, M., Bagheri, R., Fabrication and characterization of a starch-based nanocomposite scaffold with highly porous and gradient structure for bone tissue engineering. *Biomed. Phys. Eng. Express*, 4, 1–25, 2018.
21. Chakraborty, S., Ponrasu, T., Chandel, S., Dixit, M., Muthuvijayan, V., Reduced graphene oxide-loaded nanocomposite scaffolds for enhancing angiogenesis in tissue engineering applications. *R. Soc. Open Sci.*, 5, 1–14, 2018.
22. Kaur, T. and Thirugnanam, A., Tailoring *in vitro* biological and mechanical properties of polyvinyl alcohol reinforced with threshold carbon nanotube concentration for improved cellular response. *RSC Adv.*, 6, 39982–39992, 2016.
23. Millon, L.E., Oates, C.J., Wan, W., Compression properties of polyvinyl alcohol-bacterial cellulose nanocomposite. *J. Biomed. Mater. Res. Part B Appl. Biomater.*, 90, 922–929, 2009.
24. Shamekhi, M.A., Mirzadeh, H., Mahdavi, H., Rabiee, A., Mohebbi-Kalhori, D., Eslaminejad, M.B., Graphene oxide containing chitosan scaffolds for cartilage tissue engineering. *Int. J. Biol. Macromol.*, 127, 396–405, 2019.
25. Kanimozhi, K., Khaleel Basha, S., Sugantha Kumari, V., Kaviyarasu, K., Maaza, M., *In vitro* cytocompatibility of chitosan/PVA/methylcellulose–

Nanocellulose nanocomposites scaffolds using L929 fibroblast cells. *Appl. Surf. Sci.*, 449, 574–583, 2018.

26. Mc Gann, M.J., Higginbotham, C.L., Geever, L.M., Nugent, M.J.D., The synthesis of novel pH-sensitive poly(vinyl alcohol) composite hydrogels using a freeze/thaw process for biomedical applications. *Int. J. Pharm.*, 372, 154–61, 2009.
27. Nugent, M.J.D. and Higginbotham, C.L., Preparation of a novel freeze thawed poly(vinyl alcohol) composite hydrogel for drug delivery applications. *Eur. J. Pharm. Biopharm.*, 67, 377–386, 2007.
28. Nugent, M.J.D., Hanley, A., Tomkins, P.T., Higginbotham, C.L., Investigation of a novel freeze-thaw process for the production of drug delivery hydrogels. *J. Mater. Sci. Mater. Med.*, 16, 1149–1158, 2005.
29. Kalita, H., Pal, P., Dhara, S., Pathak, A., Fabrication and characterization of polyvinyl alcohol/metal (Ca, Mg, Ti) doped zirconium phosphate nanocomposite films for scaffold-guided tissue engineering application. *Mater. Sci. Eng. C*, 71, 363–371, 2017.
30. Gonzalez, J.S., Maiolo, A.S., Hoppe, C.E., Alvarez, V.A., Composite gels based on poly(vinyl alcohol) for biomedical uses. *Proc. Mater. Sci.*, 1, 483–490, 2012.
31. Hajeeassa, K.S., Hussein, M.A., Anwar, Y., Tashkandi, N.Y., Al-Amshany, Z.M., Nanocomposites containing polyvinyl alcohol and reinforced carbon-based nanofiller: A super effective biologically active material. *Nanobiomedicine*, 5, 1–12, 2018.
32. Gautam, A. and Ram, S., Preparation and thermomechanical properties of Ag-PVA nanocomposite films. *Mater. Chem. Phys.*, 119, 266–271, 2010.
33. Jarosz, M., Kapusta-Kołodziej, J., Pawlik, A., Syrek, K., Sulka, G.D., Drug delivery systems based on titania nanostructures, in: *Nanostructures for Drug Delivery*, Elsevier Inc., Amsterdan, 2017.
34. Mei, L., Zhang, Z., Zhao, L., Huang, L., Yang, X.L., Tang, J., Feng, S.S., Pharmaceutical nanotechnology for oral delivery of anticancer drugs. *Adv. Drug Deliv. Rev.*, 65, 880–890, 2013.
35. Tiwari, G., Tiwari, R., Sriwastawa, B., Bhati, L., Pandey, S., Pandey, P., Bannerjee, S.K., Drug delivery systems: An updated review. *Int. J. Pharm. Investig.*, 2, 2–11, 2012.
36. Zhu, Y.J. and Chen, F., pH-responsive drug-delivery systems. *Chem. Asian J.*, 10, 284–305, 2015.
37. Secret, E. and Andrew, J.S., Chapter 9. Enzyme-responsive drug delivery systems, in: *Stimuli-Responsive Drug Delivery System*, pp, 209–231, 2018.
38. Almeida, A., Macedo, H., das Neves, J., Sarmento, B., Chapter 4. Thermo-responsive nanomedicines for drug delivery in the gastrointestinal tract, in: *Stimuli-Responsive Drug Delivery Systems*, pp. 83–108, 2018.

39. Dai, L., Liu, J., Luo, Z., Li, M., Cai, K., Tumor therapy: Targeted drug delivery systems. *J. Mater. Chem. B*, 4, 6758–6772, 2016.
40. Griffith, L.G., Polymeric biomaterials. *Acta Materialia*, 48, 1, 263–277, 2000. (https://www.sciencedirect.com/science/article/pii/S1359645499002992)
41. Chen, Y., McCulloch, R.K., Gray, B.N., Synthesis of albumin-dextran sulfate microspheres possessing favourable loading and release characteristics for the anticancer drug doxorubicin. *J. Control. Release*, 31, 49–54, 1994.
42. Calvo, P., Remuñan-López, C., Vila-Jato, J.L., Alonso, M.J., Chitosan and chitosan/ethylene oxide-propylene oxide block copolymer nanoparticles as novel carriers for proteins and vaccines. *Pharm. Res.*, 14, 1431–1436, 1997.
43. Fresta, M., Puglisi, G., Giammona, G., Cavallaro, G., Micali, N., Furneri, P.M., Pefloxacine mesilate-and ofloxacin-loaded polyethylcyanoacrylate nanoparticles: Characterization of the colloidal drug carrier formulation. *J. Pharm. Sci.*, 84, 895–902, 1995.
44. Roy, S.K. and Prabhakar, B., Bioadhesive polymeric platforms for transmucosal drug delivery systems-a review. *Trop. J. Pharm. Res.*, 9, 91–104, 2010.
45. Sharma, K., Singh, V., Arora, A., Natural biodegradable polymers as matrices in transdermal drug delivery. *Int. J. Drug Dev. Res.*, 3, 85–103, 2011.
46. Satturwar, P.M., Fulzele, S.V., Dorle, A.K., Biodegradation and *in vivo* biocompatibility of rosin: A natural film-forming polymer. *AAPS PharmSciTech*, 4, 434–439, 2004.
47. Tang, Y., Zhao, Y., Li, Y., Du, Y., A thermosensitive chitosan/poly(vinyl alcohol) hydrogel containing nanoparticles for drug delivery. *Polym. Bull.*, 64, 791–804, 2010.
48. Arias, J.L., López-Viota, M., Sáez-Fernández, E., Ruiz, M.A., Delgado, Á. V., Engineering of an antitumor (core/shell) magnetic nanoformulation based on the chemotherapy agent ftorafur. *Colloids Surf. A Physicochem. Eng. Asp.*, 384, 157–163, 2011.
49. Ahmadian, Y., Bakravi, A., Hashemi, H., Namazi, H., Synthesis of polyvinyl alcohol/CuO nanocomposite hydrogel and its application as drug delivery agent. *Polym. Bull.*, 76, 1967–1983, 2019.
50. Wong, B.S., Yoong, S.L., Jagusiak, A., Panczyk, T., Ho, H.K., Ang, W.H., Pastorin, G., Carbon nanotubes for delivery of small molecule drugs. *Adv. Drug Deliv. Rev.*, 65, 1964–2015, 2013.
51. Maeda, H., The enhanced permeability and retention (EPR) effect in tumor vasculature: The key role of tumor-selective macromolecular drug targeting. *Adv. Enzyme Regul.*, 41, 189–207, 2001.
52. Bai, H., Li, C., Wang, X., Shi, G., A pH-sensitive graphene oxide composite hydrogel. *Chem. Commun.*, 46, 2376–2378, 2010.

53. Mahmoudi, M., Sant, S., Wang, B., Laurent, S., Sen, T., Superparamagnetic iron oxide nanoparticles (SPIONs): Development, surface modification and applications in chemotherapy. *Adv. Drug Deliv. Rev.*, 63, 24–46, 2011.
54. Arruebo, M., Fernández-pacheco, R., Ibarra, M.R., Santamaría, J., Magnetic nanoparticles for drug delivery. *Nanotoday*, 2, 22–32, 2007.
55. Mehra, N.K. and Palakurthi, S., Interactions between carbon nanotubes and bioactives: A drug delivery perspective. *Drug Discov. Today*, 21, 585–597, 2016.
56. Irmak, E.T., Carbon nanotube based polyvinylalcohol-polyvinylpyrolidone nanocomposite hydrogels for controlled drug delivery applications, Anadolu Univ. *J. Sci. Technol. A Appl. Sci. Eng.*, 18, 1–1, 2017.
57. Mu, S., Li, G., Liang, Y., Wu, T., Ma, D., Hyperbranched polyglycerol-modified graphene oxide as an efficient drug carrier with good biocompatibility. *Mater. Sci. Eng. C*, 78, 639–646, 2017.
58. Huang, H.Y., Hu, S.H., Chian, C.S., Chen, S.Y., Lai, H.Y., Chen, Y.Y., Self-assembling PVA-F127 thermosensitive nanocarriers with highly sensitive magnetically-triggered drug release for epilepsy therapy *in vivo*. *J. Mater. Chem.*, 22, 8566–8573, 2012.
59. Chiang, C.S., Tseng, Y.H., Liao, B.J., Chen, S.Y., Magnetically targeted nanocapsules for PAA-cisplatin-conjugated cores in PVA/SPIO shells via surfactant-free emulsion for reduced nephrotoxicity and enhanced lung cancer therapy. *Adv. Healthc. Mater.*, 4, 1066–1075, 2015.
60. Robinson, M., Steed, D., Franz, M., Foreword, current problems surgery. 48, 140, 2001.
61. Amrani, S., Halimi, Y., Tahiri, M., Composite materials using expanded perlite as a charge and plastic wastes as reinforcement, elaboration and properties. *GSTF Int. J. Chem. Sci.*, 1, 47–68, 2014.
62. Daunton, C., Kothari, S., Smith, L., Steele, D., A history of materials and practices for wound management. *Wound Pract. Res. Aust. J. Wound Manage.*, 20, 174–186, 2012.
63. Martin, L., Wilson, C.G., Koosha, F., Tetley, L., Gray, A.I., Senel, S., Uchegbu, I.F., The release of model macromolecules may be controlled by the hydrophobicity of palmitoyl glycol chitosan hydrogels. *J. Control. Release*, 80, 87–100, 2002.
64. Gwon, H., Lim, Y., An, S., Youn, M., Han, S., Characterization of PVA/glycerin hydrogels using γ-irradiation for advanced wound dressings. *Korean J. Chem. Eng.*, 26, 1686–1688, 2009.
65. Kamoun, E.A., Kenawy, E.R.S., Chen, X., A review on polymeric hydrogel membranes for wound dressing applications: PVA-based hydrogel dressings. *J. Adv. Res.*, 8, 217–233, 2017.

66. Bhowmick, S., Mohanty, S., Koul, V., Fabrication of transparent quaternized PVA/silver nanocomposite hydrogel and its evaluation as an antimicrobial patch for wound care systems. *J. Mater. Sci. Mater. Med.*, 66, 0–1, 2016.
67. Arıkan, H.K. and Solak, H.H., Propolis extract-PVA nanocomposites of textile design: Antimicrobial effect on gram positive and negative bacterias. *Int. J. Second. Metab.*, 4, 218–224, 2017.
68. Vimala, K., Mohan, Y.M., Varaprasad, K., Narayana, N., Fabrication of curcumin encapsulated chitosan-PVA silver nanocomposite films for improved antimicrobial activity. *J. Biomater. Nanobiotechnol.*, 1, 55–64, 2011.
69. Tarun, K. and Gobi, N., Calcium alginate/PVA blended nano fibre matrix for wound dressing. *Indian J. Fibre Text. Res.*, 37, 127–132, 2012.
70. Rafienia, M., Saberi, A., Poorazizi, E., A novel fabrication of PVA/alginate-bioglass electrospun for biomedical engineering application. *Nanomedicine,* 4, 152–163, 2017.
71. Noori, S., Kokabi, M., Hassan, Z.M., Nanoclay enhanced the mechanical properties of poly(vinyl alcohol)/Chitosan/montmorillonite nanocomposite hydrogel as wound dressing. *Proc. Mater. Sci.*, 11, 152–156, 2015.
72. Kokabi, M., Sirousazar, M., Hassan, Z.M., PVA-clay nanocomposite hydrogels for wound dressing. *Eur. Polym. J.*, 43, 773–781, 2007.
73. Ray, S.S. and Okamoto, M., Polymer/layered silicate nanocomposites : A review from preparation to processing. *Prog. Polym. Sci.,* 28, 1539–1641, 2003.
74. Ali, M.H.M., Kahder, M.M., Al-saad, K.A., Al-meer, S., Properties of nanoclay PVA composites materials. *QScience Connect,* 2013, 1, 1–9, 2013.
75. George, J., Sajeevkumar, V.A., Ramana, K.V., Sabapathy, S.N., Siddaramaiah, Augmented properties of PVA hybrid nanocomposites containing cellulose nanocrystals and silver nanoparticles. *J. Mater. Chem.*, 22, 22433–22439, 2012.
76. Osorio, M., Velásquez-Cock, J., Restrepo, L.M., Zuluaga, R., Gañán, P., Rojas, O.J., Ortiz-Trujillo, I., Castro, C., Bioactive 3D-shaped wound dressings synthesized from bacterial cellulose: Effect on cell adhesion of polyvinyl alcohol integrated *in situ*. *Int. J. Polym. Sci.*, 2017, 2017.
77. Khalilipour, A. and Paydayesh, A., Characterization of polyvinyl alcohol/ ZnO nanocomposite hydrogels for wound dressings. *J. Macromol. Sci. Part B.*, 58, 371–384, 2019.
78. Hassan, H.S., Elkady, M.F., Hafez, E.E., Salama, E., Novel antibacterial zinc oxide polymeric nanocomposite membrane as wound dress. *Nanosci. Nanotechnol. Asia*, 7, 62–72, 2017.
79. Wang, X., Chang, J., Wu, C., Bioactive inorganic/organic nanocomposites for wound healing. *Appl. Mater. Today*, 11, 308–319, 2018.

80. Wang, P., Lu, Q., Li, J., *In situ* formation of palladium nanoparticles inside the pore channels of ordered mesoporous silica. *Catal. Lett.*, 131, 444–450, 2009.
81. Roper, W.L., *Toxicological profile for silver*. Agency for Toxic Substances and Disease Registry. U.S. Public Health Service, 1990.
82. Jatoi, A.W., Ogasawara, H., Kim, I.S., Ni, Q.Q., Polyvinyl alcohol nanofiber based three phase wound dressings for sustained wound healing applications. *Mater. Lett.*, 241, 168–171, 2019.
83. Ishtiaq, S., Ullah, A., Ali, K., Attaullah, M., Khan, H., Ali, H., Amjad, M., Tahir, M., Javed, M., Ghramh, H.A., Adgaba, N., Kanta, C., Composition and functional properties of propolis (bee glue): A review. *Saudi J. Biol. Sci.*, 6 1695–1703, 2018.
84. Sforcin, J.M., Bankova, V., Kuropatnicki, A.K., Medical benefits of honeybee products. *Evid. Based Complement Alternat. Med.*, 2017, 2–4, 2017.
85. Toreti, V.C., Sato, H.H., Pastore, G.M., Park, Y.K., Recent progress of propolis for its biological and chemical compositions and its botanical origin. *Evid. Based Complement Alternat. Med.*, 2013, 13, 2013.
86. Sforcin, J.M., Biological properties and therapeutic applications of propolis. *Phytother. Res.*, 905, 894–905, 2016.
87. Rajpara, S., Wilkinson, M.S., King, C.M., Gawkrodger, D.J., English, J.S.C., Statham, B.N., Green, C., Sansom, J.E., Chowdhury, M.M.U., Horne, H.L., Ormerod, A.D., The importance of propolis in patch testing–A multicentre survey. *Contact Dermatitis*, 61, 5, 287–290, 2009.
88. Kamel, N.A., Soliman, A.A.F., Rozik, N.N., Abd-Elmessieha, S.L., Biophysical investigation of curcumin based nanocomposite for wound dressing application. *J. Appl. Pharm. Sci.*, 8, 35–44, 2018.
89. Sirousazar, M., Kokabi, M., Hassan, Z.M., Bahramian, A.R., Dehydration kinetics of polyvinyl alcohol nanocomposite hydrogels containing Na-montmorillonite nanoclay. *Sci. Iran.*, 18, 780–784, 2011.
90. Koosha, M., Mirzadeh, H., Shokrgozar, M.A., Farokhi, M., Nanoclay-reinforced electrospun chitosan/PVA nanocomposite nanofibers for biomedical applications. *RSC Adv.*, 5, 10479–10487, 2015.
91. Watanabe, K., Tabuchi, M., Ishikawa, A., Takemura, H., Tsuchida, T., Morinaga, Y., Yoshinaga, F., Acetobacter xylinum mutant with high cellulose productivity and an ordered structure. *Biosci. Biotechnol. Biochem.*, 62, 1290–1292, 1998.
92. Gelinsky, M. and Heinemann, S., Nanocomposites for tissue engineering, in: *Nanocomposites for Life Sciences*, vol. 8, pp. 405–434, 2010.
93. O'Brien, F.J., Biomaterials & scaffolds for tissue engineering. *Mater. Today*, 14, 88–95, 2011.
94. Shuai, C., Feng, P., Gao, C., Shuai, X., Nanocomposite scaffolds for tissue engineering. *RSC Adv.*, 5, 25416–25423, 2015.

95. Kashyap, S., Pratihar, S.K., Behera, S.K., Strong and ductile graphene oxide reinforced PVA nanocomposites. *J. Alloys Compd.*, 684, 254–260, 2016.
96. Prasadh, S., Suresh, S., Wong, R., Osteogenic potential of graphene in bone tissue engineering scaffolds. *Mater. (Basel)*, 11, 1430, 2018.
97. Shokrgozar, M.A., Mottaghitalab, F., Mottaghitalab, V., Farokhi, M., Fabrication of porous chitosan/poly(vinyl alcohol) reinforced single-walled carbon nanotube nanocomposites for neural tissue engineering. *J. Biomed. Nanotechnol.*, 7, 2, 276–284, 2011.
98. Liao, H., Qi, R., Shen, M., Cao, X., Guo, R., Zhang, Y., Shi, X., Improved cellular response on multiwalled carbon nanotube-incorporated electrospun polyvinyl alcohol/chitosan nanofibrous scaffolds. *Colloids Surf. B Biointerfaces*, 84, 528–535, 2011.
99. Brown, J., Kumbar, S., Banik, B., *Bio-Instructive Scaffolds for Musculoskeletal Tissue Engineering and Regenerative Medicine*, Elsevier Science, Amsterdam, 2016.
100. Abdal-Hay, A., Kim, C.I., Lim, J.K., An *in situ* hydrothermal fabrication process of poly(vinyl alcohol)/apatite-like nanocomposites with improved thermal and mechanical properties. *Ceram. Int.*, 40, 4995–5000, 2014.
101. Bao, L., Hong, F.F., Chen, L., Li, X., Tang, J., Potential of PVA-doped bacterial nano-cellulose tubular composites for artificial blood vessels. *J. Mater. Chem. B*, 3, 8537–8547, 2015.
102. Mohanapriya, S., Mumjitha, M., PurnaSai, K., Raj, V., Fabrication and characterization of poly(vinyl alcohol)-TiO2 nanocomposite films for orthopedic applications. *J. Mech. Behav. Biomed. Mater.*, 63, 141–156, 2016.
103. Cai, D. and Song, M., Recent advance in functionalized graphene/polymer nanocomposites. *J. Mater. Chem.*, 20, 7906–7915, 2010.
104. Veeriah, V., Suresh, K.I., Chatterjee, S., Patra, C.R., Barui, A.K., Mukherjee, S., Sriram, P., Nethi, S.K., Graphene oxides show angiogenic properties. *Adv. Healthc. Mater.*, 4, 1722–1732, 2015.
105. Shin, S.R., Tamayol, A., Khoshakhlagh, P., Li, Y.-C., Zhang, Y.S., Khademhosseini, A., Akbari, M., Jang, H.L., Nasajpour, A., Graphene-based materials for tissue engineering. *Adv. Drug Deliv. Rev.*, 105, 255–274, 2016.
106. Freeman, J.W. and Browe, D.P., Bio-instructive scaffolds for skeletal muscle regeneration: Conductive materials, Elsevier Inc, Amsterdam, 2016.
107. Mehta, M. and Subramani, K., Nanodiagnostics in microbiology and dentistry, First, in: *Emerging Nanotechnologies in Dentistry*, Elsevier Inc, Amsterdam, 2012.
108. Weiner, S. and Wagner, H.D., The material bone: Structure-mechanical function relations. *Annu. Rev. Mater. Sci.*, 28, 271–298, 2002.
109. Acton, Q.A., *Apatites—Advances in Research and Application*, 2012 Edition: ScholarlyBrief, Scholarly Editions, Atlanta US, 2012.

110. Zhou, H. and Lee, J., Nanoscale hydroxyapatite particles for bone tissue engineering. *Acta Biomater.*, 7, 2769–2781, 2011.
111. Wang, X., Grogan, S.P., Rieser, F., Winkelmann, V., Maquet, V., La Berge, M., Mainil-Varlet, P., Tissue engineering of biphasic cartilage constructs using various biodegradable scaffolds: An *in vitro* study. *Biomaterials*, 25, 3681–3688, 2004.
112. Pina, S., Oliveira, J.M., Reis, R.L., Natural-based nanocomposites for bone tissue engineering and regenerative medicine: A review. *Adv. Mater.*, 27, 1143–1169, 2015.
113. Parsa, P., Paydayesh, A., Davachi, S.M., Investigating the effect of tetracycline addition on nanocomposite hydrogels based on polyvinyl alcohol and chitosan nanoparticles for specific medical applications. *Int. J. Biol. Macromol.*, 121, 1061–1069, 2019. https://doi.org/10.1016/j.ijbiomac.2018.10.074.
114. Enayati, M.S., Behzad, T., Sajkiewicz, P., Rafienia, M., Bagheri, R., Ghasemi-Mobarakeh, L., Kolbuk, D., Pahlevanneshan, Z., Bonakdar, S.H., Development of electrospun poly(vinyl alcohol)-based bionanocomposite scaffolds for bone tissue engineering. *J. Biomed. Mater. Res. A*, 106, 1111–1120, 2018.
115. Koosha, M., Raoufi, M., Moravvej, H., One-pot reactive electrospinning of chitosan/PVA hydrogel nanofibers reinforced by halloysite nanotubes with enhanced fibroblast cell attachment for skin tissue regeneration. *Colloids Surf. B Biointerfaces*, 179, 270–279, 2019.
116. Tamburaci, S., Cecen, B., Ustun, O., Ergur, B.U., Havitcioglu, H., Tihminlioglu, F., Production and characterization of a novel bilayer nanocomposite scaffold composed of chitosan/si-nHap and zein/POSS structures for osteochondral tissue regeneration. *ACS Appl. Bio Mater.*, 4, 1440–1455, 2019.
117. Discher, D.E., Janmey, P., Wang, Y.L., Tissue cells feel and respond to the stiffness of their environment. *Science*, 10, 1139–1143, 2005.
118. Wang, H., Morales, R.T.T., Cui, X., Huang, J., Qian, W., Tong, J., Chen, W., A photoresponsive hyaluronan hydrogel nanocomposite for dynamic macrophage immunomodulation. *Adv. Healthc. Mater.*, 8, 1–9, 2019.
119. Deller, R.C., Richardson, T., Richardson, R., Bevan, L., Zampetakis, I., Scarpa, F., Perriman, A.W., Artificial cell membrane binding thrombin constructs drive *in situ* fibrin hydrogel formation. *Nat. Commun.*, 10, 1887, 2019.
120. Nguyen, A.T., Sathe, S.R., Yim, E.K.F., From nano to micro: Topographical scale and its impact on cell adhesion, morphology and contact guidance. *J. Phys. Condens. Matter*, 28, 183001, 2016.
121. Harley, B.A.C., Kim, H.-D., Zaman, M.H., Yannas, I.V., Lauffenburger, D.A., Gibson, L.J., Microarchitecture of three-dimensional scaffolds influences cell migration behavior via junction interactions. *Biophys. J.*, 95, 4013–4024, 2008.

122. Petersen, A., Joly, P., Bergmann, C., Korus, G., Duda, G.N., The impact of substrate stiffness and mechanical loading on fibroblast-induced scaffold remodeling. *Tissue Eng.*, 18, 1804–1817, 2012.
123. Rnjak-Kovacina, J., Wise, S.G., Li, Z., Maitz, P.K.M., Young, C.J., Wang, Y., Weiss, A.S., Tailoring the porosity and pore size of electrospun synthetic human elastin scaffolds for dermal tissue engineering. *Biomaterials*, 32, 6729–6736, 2011.
124. Efraim, Y., Schoen, B., Zahran, S., Davidov, T., Vasilyev, G., Baruch, L., Zussman, E., Machluf, M., 3D structure and processing methods direct the biological attributes of ECM-based cardiac scaffolds. *Sci. Rep.*, 9, 5578, 2019.
125. Zhou, D., Dong, Q., Liang, K., Xu, W., Zhou, Y., Xiao, P., Photocrosslinked methacrylated poly(vinyl alcohol)/hydroxyapatite nanocomposite hydrogels with enhanced mechanical strength and cell adhesion. *J. Polym. Sci. A Polym. Chem.*, 57, 1–8, 2018.
126. Fujie, T., Shi, X., Ostrovidov, S., Liang, X., Nakajima, K., Chen, Y., Wu, H., Khademhosseini, A., Spatial coordination of cell orientation directed by nanoribbon sheets. *Biomaterials*, 53, 86–94, 2015.
127. Liang, S., Zhang, Y., Wang, H., Xu, Z., Chen, J., Bao, R., Tan, B., Cui, Y., Fan, G., Wang, W., Paintable and rapidly bondable conductive hydrogels as therapeutic cardiac patches. *Adv. Mater.*, 30, 1704235, 2018.
128. Tognato, R., Armiento, A.R., Bonfrate, V., Levato, R., Malda, J., Alini, M., Eglin, D., Giancane, G., Serra, T., Nanocomposite biomaterials: A stimuli-responsive nanocomposite for 3D anisotropic cell-guidance and magnetic soft robotics. *Adv. Funct. Mater.*, 29, 1970052, 2019.

8

Hybrid Interpolymeric Complexes

Igor Prosanov

Institute of Solid State Chemistry and Mechanochemistry, Siberian Branch of Russian Academy of Sciences, Novosibirsk, Russia

Abstract

Hybrid interpolymeric complexes (HICs) compose a new class of materials. Such complex consists of inorganic and organic complexing polymers bonded one with another by noncovalent interactions like DNA double helix. Inorganic chain polymers stabilized in organic matrix can be produced from HICs through solid state synthethis. These materials can be used in different applications.

Keywords: Polyvinyl alcohol, hybrid interpolymeric complex, polymeric complex, inorganic polymers

8.1 Introduction

8.1.1 Historical Overview

Hybrid interpolymeric complexes (HICs) are a new class of materials. Such complex consists of inorganic and organic polymers bonded one with another by noncovalent interactions. Earlier, organic interpolymeric complexes were known. The most prominent representative of such organic complexes is deoxyribonucleic acid (DNA). It is known that DNA complex includes two polymeric chains bonded through complementary nucleotide groups. HIC can be viewed to some extent as an analog of DNA where one component is inorganic polymer. The existence of HICs as a new class of materials was predicted for the first time in 2013 [1]. Earlier, only two representatives of this class were discussed. One

Email: prosanov@mail.ru

Visakh P. M. and Olga B. Nazarenko (eds.) Polyvinyl Alcohol-Based Biocomposites and Bionanocomposites, (205–252) © 2023 Scrivener Publishing LLC

of them is a complex of copper hydroxide with polyvinyl alcohol (PVA) (–CH_2–CHOH–)n [2]. Then, well-known complex of starch or PVA with iodine also can be attributed to such materials. It is supposed that in this complex, iodine molecules I_2 form a linear chain inside a helix of polysaccharide or PVA. Within iodine chain I_2 single molecules apparently can be still recognized. Different structures of this complex are also discussed in researches [3]. We do not know publications where other HICs were discussed before 2013.

Some researchers discussed relative structures. For instance, in Wenbin *et al.* [4], it was proposed that zinc hydroxide molecules can be adsorbed on a PVA chain.

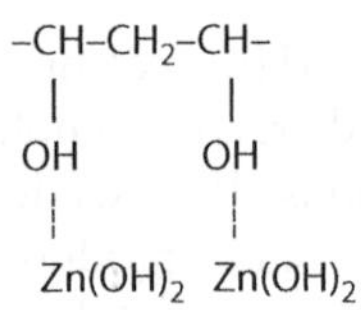

(1)

It is natural to suggest that such way absorbed $Zn(OH)_2$ molecules actually will interact with each other and form inorganic polymeric chain bonded with PVA chain.

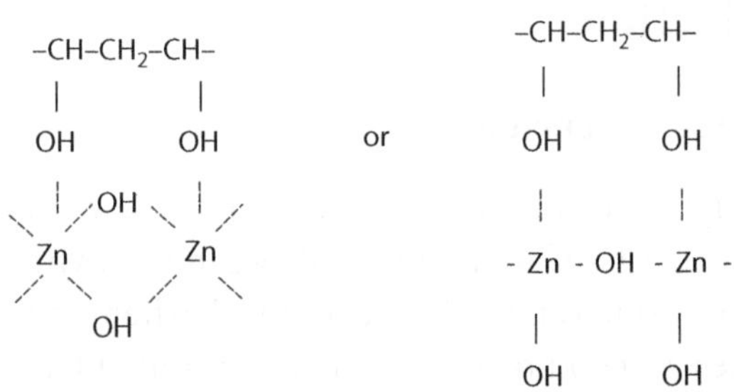

(2)

Here we see two polymeric chains. One of them is organic complexing polymer — polyvinyl alcohol, and another one is inorganic substance — zinc hydroxide in polymeric form. Both chains are interconnected with coordinative bonds forming hybrid interpolymeric complex. The peculiarity of inorganic component is a coordinative bonding within it. The formal valence of elements (2 for Zn) and chemical groups (one for OH) in this polymer differs from the number of bonds attributed to them (5 for Zn and 2 for some OH groups in our example). Therefore, such polymers are named for "coordinative polymers" (CPs). The possibility of OH group to coordinative bonds formation is well known [5]. Halide ions (Cl, Br, and I) possess of this property as well [5]. Therefore, it is natural to expect the formation of polymeric halides (like zinc halide) as components of HICs.

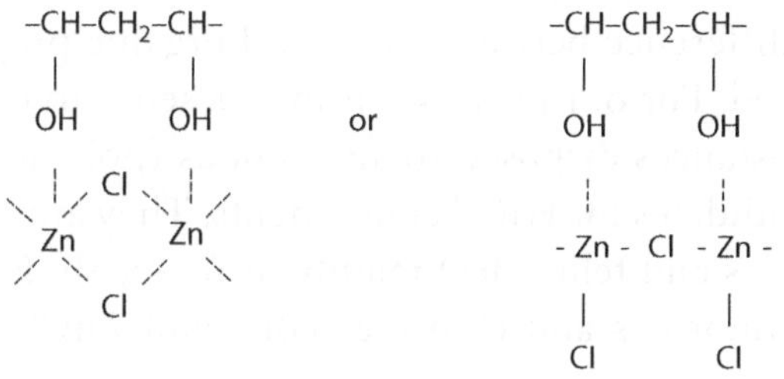

(3)

Many halides are well water soluble compounds and we believe that they can form HICs in solutions with water soluble complexing organic polymers. After drying they give solid substances. Chemical reactions in solid state have peculiarities in comparison with liquid phase reactions. In particular, the metastable nonequilibrium products can be obtained in this case due to diffusion hindering [6]. Therefore, we can hope to obtain compound (2) from (3) by using substitution reaction:

$$ZnCl_2 + NH_4OH \rightarrow Zn(OH)_2 \downarrow + NH_4Cl \quad (4)$$

in presence of PVA as a kind of solid solvent.

The experimental data and analyzes described above have led the author to the concept of HICs.

8.1.2 General Description of HICs

One distinctive component of HIC is an inorganic polymer. Therefore, it would worth to make some remarks about these polymers existence to understand HICs concept. It is commonly accepted that inorganic substances have much weaker disposition to produce polymeric form in comparison with organic one. In the final analysis, it is explained by relative nondirectivity of bonds in inorganic compounds. They exhibit rather coordinative bonds where the number of nearest atoms can differ from a formal valence of an element. Inorganic compounds in general tend to formation of crystalline structures instead of polymeric ones. There are some exceptions from this common rule like Selenium,

Sulfur and several other elements. Some inorganic compounds like silica, boron and phosphorus oxides, for example, can form peculiar polymeric structures called glasses [7]. In their structures chains of alternating Si, B or P and O atoms can be distinguished. In a whole, the existence of polymeric inorganic structures is commonly adopted but the prominent difference between them and organic polymers should also be recognized. For our purposes it makes sense to draw attention to inorganic substances disposed to amorphous (polymeric) structure formation as candidates for HICs' components. They are chalcogenides (sulfides, selenides and tellurides) mainly of B, As, Sb, P, Si or Ge and halides usually fluorides and chlorides (BeF and $ZnCl_2$ as examples) [7].

Taking into account mentioned above it is expected that HICs should include coordinative inorganic polymers. For example, the formation of PVA-$Cu(OH)_2$ HIC from complexes of PVA with single molecules of copper hydroxide can be described by the following scheme at the rising of $Cu(OH)_2$ content in a system.

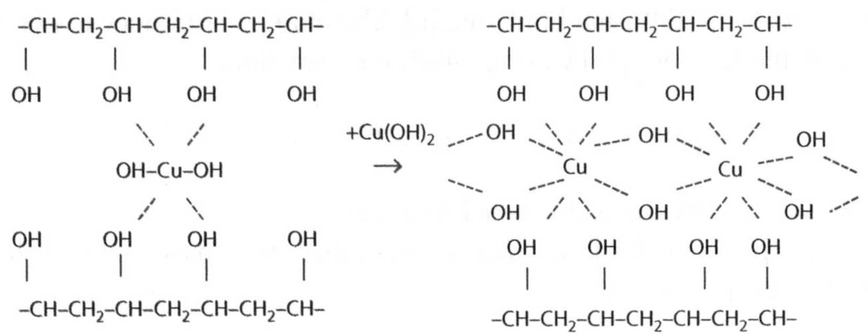

In this scheme, we see one molecule and then one chain of copper hydroxide complexing with two chains of PVA. The separate $Cu(OH)_2$ molecules can not be distinguished in this inorganic polymer. It seems that represented stoichiometry is not obligatory in this case. The complex can include one chain of copper hydroxide complexing with one or more chains of PVA. Generally speaking, stoichiometry of HICs is not strictly defined value. After thermal dehydration, this complex is expecting transforms into the complex of PVA with polymeric copper oxide CuO.

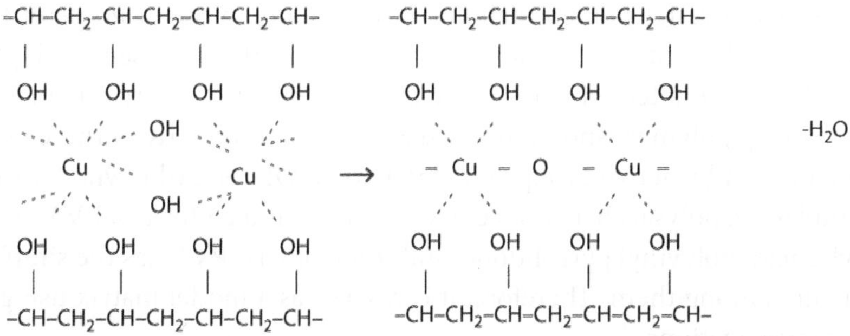

(5)

Evidently, polymeric copper oxide is metastable substance. It can be produced due to diffusion hardship in solid state. This sequence of transformations represents a common way of HICs production through solid state reactions starting from complexes of soluble compounds. In other words, solid state chemistry is a substantial part of HICs preparation.

It is interesting to note that polymeric copper oxide can be viewed also as a component of high temperature superconductors [8]. Therefore, the hope exists to find similar unusual properties at HICs.

Complex of PVA with iodine has its peculiarities. In simplest form it can be depicted as following.

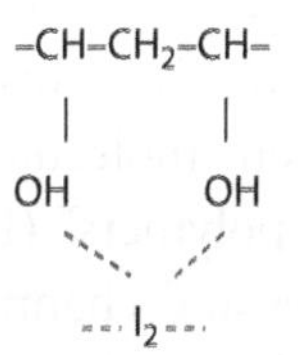

Iodine molecules I_2 can be distinguished here as building units. They interact each other and with PVA's OH groups through coordinative bonds. In more complicated case I_3^- particles can represent separate polymeric units and counterions like K^+ shall be included in a system as well. It can be concluded from given examples that distinguishing whether inorganic species exist within a complex in molecular or polymeric form is important in HICs concept.

Nowadays an active search for new materials promising for applications is continuing and different metastable substances like graphene and metal-organic frameworks (MOFs) enter the market. These substances have intention to high activity in physical and chemical processes. From the one hand, it makes them prospective for wide number of applications but from another hand restricts their stability. HICs are one more example of this concept implementation. Inorganic polymer in HIC represents a highly active form of matter, which is needed to be stabilized by organic matrix. Complexing polymers should be used as such matrix in HICs. These polymers are rich by active groups like hydroxylic OH or carboxylic OCOH. Examples are polysaccharides (cellulose, starch, and chitozan), PVA, polyacrylic acid, polyvinyl pyrrolidone, and some others. PVA has the simplest structure among them. Therefore, it can serve as a model matrix using in HICs investigations.

8.1.3 Relative Materials

It has been suggested above that HICs can be regarded as special case of complexes of inorganic molecules with complexing organic polymers. In other words, they are relative materials with some quite similar properties. Other materials relative to HICs can also be pointed. Among them are coordinative polymers (CPs) in which repeated units linked by coordinative bonds can be identified. Mentioned above MOFs are the most known kind of CPs. Hybrid crystalline compounds represent other sort of CPs [9, 10]. Now, MOFs are subjects of numerous researches due to peculiarities of their structure that gives rich possibilities for design of functional materials based on them. The main distinctive feature of MOFs is presence of pores and channels that makes them suitable for different molecules intercalation. MOFs are also called as "porous coordinative polymers" (PCPs). Other hybrid crystalline compounds have no such pores and channels. HICs differ from MOFs in use of polymeric organic ligands instead of low-molecular ones. It leads to noncrystallinity of HICs in contrast to MOFs. On the other hand, inorganic component is cohesive in HICs as opposed to MOFs. It gives a possibility to direct current passing through inorganic part of HICs and makes possible different applications based on this property like photoenergy conversion and gas sensing. There are no regular pores and channels in HICs but different known common methods can be used for pores formation in HICs as a sort of polymers. Finally, it can be stated that HICs are sort of CPs. One more class of relative materials embraces hybrid composites. They also contain inorganic substance in form of particles with different morphology in organic matrix and can possess some properties similar to properties of HICs. Such similarity is expected be higher for HICs and nanocomposites which can be viewed as intermediate species between ordinary compositions and HICs. The most substantial difference between composites and HICs is the presence/absence of phase boundaries correspondently. First materials are classified as heterogeneous structures and the second as homogeneous ones. The presence of boundaries can sufficiently change the properties of a bulk sample. This change increases with rising of boundary surface i.e. with dispersiveness increase.

Another classification of relative materials can be developed for inorganic polymers as a part of HICs. It is based on their one-dimensionality. One group in this classification consists of inorganic polymers. Some of them were mentioned in previous section. Here we can additionally mark polysiloxanes, polyphosphazenes, polysilanes, and others. Sufficient preparative difficulties exist on a way of inorganic polymers nomenclature extension and some specific methods of their overcoming have been suggested in a literature [11]. Another group of relative to HICs materials comprises different quasi 1-D systems like nanotubes, nanoribbons and nanowires. As a special case here an inorganic molecular wires can be pointed out [12].

8.1.4 To Summarize

- Concept of HICs was suggested first at 2013.
- Complexes of PVA with copper hydroxide and iodine were first examples of HICs described before their concept appearance.
- Discussion of PVA complex with zinc hydroxide caused HICs concept putting forward.
- Many prospective HICs are metastable structures. Therefore, their producing is based on solid state chemistry technique.
- Evidently, stoichiometry of HICs is not strictly defined value.
- HICs are kind of CPs.
- HICs are 1-D systems.

8.2 Production of HICs

Production of HICs can include three steps:

1. Producing of interpolymeric complex between soluble compounds.

2. Conversion of soluble inorganic compound (as a component of HIC) into insoluble one through a solid state substitution reaction.
3. Thermal transformation of HIC.

Operations 2) and 3) are not always needed.

As an example let us discuss production of interpolymeric complexes between PVA and zinc compounds. At the first stage soluble zinc salts like chloride or acetate are usually used [4, 13–16]. The final aim is obtaining of ZnO or ZnS species prospective for applications. It is known that solvated zinc ions readily bound by complexing polymers like PVA [4, 13, 14, 16, 17]. A type of anion counts for following processing. It is desirable that anions could be easily extracted from the system. Chlorides are accessible and have simple but quite hardly removable anion. A typical second step consists of interaction of zinc salt with ammonium water solution in accordance with expression (4). Obtained ammonium chloride should be eliminated from a system. It can be achieved by thorough washing. The problem is in slow diffusion of removing substance from solid phase. On the other hand, solvent like water can be adsorbed by HIC leading to its transformation into composite. Ammonium acetate can be removed by thermal decomposition at relatively low temperature (~ 100°C). In researches [4, 13, 14] substitution reaction (4) was carried out in liquid phase. In absence of PVA solid zinc hydroxide should precipitate in this case. Authors of [4, 14] believe that in presence of PVA zinc hydroxide gives complexes in molecular dispersive form (1) that is quite doubtful. It seems more likely that presence of PVA in solution rather influences zinc hydroxide particles morphology.

In [1] substitution reaction was carried out by treatment of PVA–$ZnBr_2$ complex in a form of solid film in water ammonium solution. The idea is following. Chemical reactions in solids can run otherwise than that in solutions [6]. It is conditioned by diffusion hindrance. A product of solid state reaction can take metastable structure if it is similar to the structure of reagent. This effect is known as predecessor principle. It can be illustrated by copper hydroxide dehydration [18].

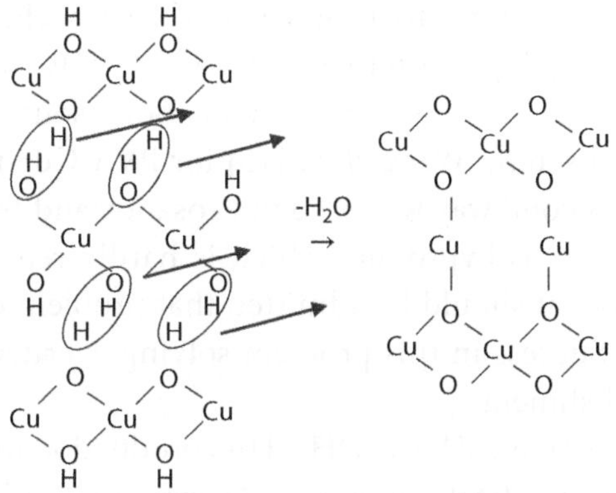

(6)

Chains of atoms can be distinguished in crystalline copper hydroxide. At dehydration they convert into chains of copper oxide. Strict correlation of initial and final structures takes place. Displacements of the atoms are minimal at such conversion. This is a clear example of the predecessor principle. The final structure is not metastable in this case but it can be metastable in common situation. The similar picture (5) we observe at PVA-$Cu(OH)_2$ HIC dehydration. In this example we suppose that polymeric structure is metastable and it is forming due to diffusion hindrance. Returning to our example with zinc hydroxide/oxide we can mark similarities and differences of both cases. One difference is that PVA-$Cu(OH)_2$ complex is stable in solution and PVA-$Zn(OH)_2$ probably is not stable. PVA-$Cu(OH)_2$ HIC can be produced directly by mixing of PVA and $Cu(OH)_2$ solutions (more precisely, a solution of ammonia $Cu(OH)_2$ complex is used) [2] while $Zn(OH)_2$ can not be dissolved. In other words, producing of PVA-$Cu(OH)_2$ HIC consists of only step 1) while producing of PVA-$Zn(OH)_2$ HIC consists of steps 1) and 2).

Using of water ammonia solution in step 2) of PVA-$Zn(OH)_2$ HIC production is not good idea because PVA absorbs water. It leads to diffusion hindrance elimination and formation of composite instead of HIC. As alternatives one can suggest treatment of solid specimen in ammonia vapor or in alcohol alkaline solution. Alcohol is not absorbed well by PVA but use of alkaline needs remove reaction byproduct. The similar problem appears at trying to produce HICs with chalcogenides. For example, if we are going to synthesize PVA-CdS HIC we can use treatment of specimen in gaseous H_2S or in alcoholic solution of Na_2S. The first method was used in Wang *et al.* [15] for PVA-CdS nanocomposites production. The second way was used in Prosanov and Bulina [19]. NaCl was produced there as byproduct which is hardly removable from the system. In a whole, it should be admitted that problem exists with stage 2) carrying out. Success in this problem solving is crucial for the HIC's concept accomplishment.

Absence of step 2) in PVA-$Cu(OH)_2$ HIC production makes this system valuable for HICs investigations as a model case. As another such a model system PVA-$B(OH)_3$ HIC can be regarded [20] due to sufficient solubility of boric acid. This system has its own peculiarities. The product of PVA and boric acid reaction precipitates from solution. It is explained by quite strong bonding of boric acid with more than one PVA molecule [21]. Such precipitation can be prevented by using of diluted (~ 1 %) PVA solutions. Several other inorganic compounds are known which also precipitate PVA complex. It was suggested that in some cases it is caused by chemical bonding [22].

Step 3) is needed for transformation of HICs with hydroxides into HICs with oxides with taking into account that the latter are viewed as more promising for applications. Some remarks should be made according to this step. First of all, the temperature of treatment should not be higher than approximately 200°C when PVA begins to destruct (in vacuum). At air PVA destruction begins at temperature near 100°C. Additives can lower destruction temperature. Therefore, few HICs are suited for thermal transformations. In particular zinc hydroxide and copper hydroxide have low temperatures of dehydration (40°C and 150°C correspondently). But, it is necessary to take into account that dehydration temperatures can be higher for HICs in comparison with crystalline hydroxides. For example, one can see from the scheme (6) that OH group of one chain interacts with nearest H atom of other chain forming H_2O molecule to be detached. There no such adjacent chains in PVA-$Cu(OH)_2$ HIC (scheme (5)). In this case OH group of one chain interacts with nearest H atom of the same chain forming H_2O molecule to be detached. Activation energy of the reaction is expected be higher in the last system. Therefore, temperature of dehydration is expected be higher as well. The temperature of PVA destruction is not limiting factor if a HIC producing is not the aim of synthesis.

High temperature treatment is used for producing of oxide nanorods and nanowires [4, 13].

8.2.1 To Summarize

- Production of HICs is started from solutions of components.
- Metastable HICs can be produced from predecessors through solid state conversion.

8.3 Structure of Hybrid Interpolymeric Complexes

8.3.1 General Description of Experimental Methods and Computations

HICs have noncrystalline structure. Therefore, X-ray diffraction (XRD) is not effective method of their investigation. There are different types of noncrystalline materials: glasses, polymers, colloids, solutions, etc. Existence of various noncrystalline structures is known as polymorphism [7]. Determining of the type of noncrystalline structure is not easy task. We use correlations between experimental data and computations as a general approach to this problem. Such method has been discussed in literature [23]. Different noncrystalline structures can be distinguished by means of vibration spectroscopy. For example, Raman spectra of amorphous silica and germanium represent a kind of phonon density of states function without distinct peaks [24]. On the other hand, spectra of polymeric substances like amorphous sulphur and selenium possess of quite sharp peaks. Therefore, considerable attention in HICs investigation is devoted to the use of vibration spectroscopy. Typical oscillation frequencies of inorganic compounds are lower than 500 cm^{-1} which is near of inferior limit of usual IR spectrometers. Therefore, Raman spectroscopy is more convenient technique for HICs investigation because it operates in needed spectral range. It is reasonable to expect that HICs vibration spectra will represent superposition of slightly changed PVA spectrum and spectra of inorganic component. Interaction between organic and inorganic polymers should cause changes in PVA bands positions. Therefore, these observed changes can be regarded as an indicator of supposed interaction. Positions of bands of inorganic components are expected not very different from positions of bands in spectra of crystalline form. Hereinafter they will be meant as

HICs' vibration bands by default. An analysis of these bands gives important information about structure of inorganic component. It allows distinguishing between polymeric, molecular and crystalline form.

Optical spectroscopy in ultraviolet (UV) and visible (Vis) ranges combined with theoretical conceptions gives information concerning electronic transitions and band structure of solids [25]. Sometimes, it is hard to prepare solid samples with needed thickness for the investigation of their transparency in UV region due to their high optical density in this range. Therefore, reflection technique is usually used for observation of electronic transitions. This method has its drawbacks. For investigation of electronic spectra of molecules their quite diluted solutions or vapors are commonly used for transmittance or absorbance measurement. HICs are interesting objects for electronic transitions investigations. They represent a kind of intermediate between solids and molecules. Quite thin films apt for transmittance method application can be easily prepared in this case. It allows more precise observing of electronic effects at solids formation.

Nuclear magnetic resonance (NMR), electron spin resonance (ESR), Mossbauer spectroscopy, low-angle X-ray diffraction, extended X-ray adsorption fine structure (EXAFS), X-ray adsorption near-edge structure (XANES) are also structure-sensitive methods which can be used. Due to author specialization, optical spectroscopy (IR, Raman, and UV-Vis) is used preferentially in his researches. Microscopy gives information about microstructure of samples. It serves as important addition to the above mentioned methods sensitive to atomic structure. Samples can exhibit different types of micromorphology: fibrous or grained with variations or be homogeneous. Combined with XRD pattern, microimages allow understand better processes governing material formation. Picnometry also should be pointed as structure sensitive method. Density of a simple binary composition can be easily calculated from the densities of matrix and filler and their ratio. If measured density of a sample differs notably from calculated one it can mean that significant interaction between two substances takes place and we cannot regard a sample like a mixture of two phases.

By means of XRD analyzes presence of crystalline phase can be determined in material. In a typical investigation a set of samples with different content of inorganic component were prepared. Then, their XRD patterns were examined. Samples with high content of inorganic component exhibit sharp peaks of crystalline phase. XRD patterns of samples with low ratio of inorganic component have smooth shape. Ratio of components when change of XRD patterns shape occurs corresponds to stoichiometry of investigated HIC.

For computation purposes we used different models: 1) Single molecules. Some of them are unknown in individual form. 2) Segments of PVA chain from 3.5 to 11.5 units. Half-integer values provide symmetry of both ends of PVA segments and minimize number of atoms in a system. 3) Dimers, chains and bulk groups up to 7 inorganic molecules. 4) One and more inorganic molecules adsorbed on PVA segment. 5) Inorganic polymeric segments up to 8 units complexed with PVA fragment. We suppose that optimized segments can serve as quite exact representatives of long polymeric chains and HICs. The bonding energies of different structures were calculated as the difference of electronic energies of these structures and their components.

Density Functional Theory (DFT) was used for obtaining HICs ground state optimized geometries, analytical vibrational frequencies, and thermochemical data (at p = 1.00 atm and T = 298.15 K). The calculations were run with the Becke three-parameter Lee–Yang–Parr (B3LYP) hybrid exchange-correlation functional [26] coupled with three basis sets: 3-21G, 6-31G, and Def2TZVP. An excitation energies and oscillator strengths in UV-Vis range were calculated at two different levels of theory: 1) The single-excitation configuration interaction (CIS) and 2) Time-Dependent DFT (TD-DFT) [27]. No remarkable difference was found for both cases. Gaussian 09 package [28] was used for calculation of all described HICs.

8.3.2 Halides of Second Group Elements as HICs Components

Halides (chlorides, bromides and iodides) of second group elements are well soluble substances. They are ideal model objects for HICs investigation due to solubility, simplicity of anion and bivalence of cations. High solubility means that substance readily forms nonvalent bonds with hydroxyl containing molecules instead of constitution of separate phase. Simplicity of anion means that there are no intrinsic oscillations which can complicate IR and Raman spectra. It seems that bivalent elements and their compounds are more disposing for polymeric chains formation. But it is rather true for covalent bonding. Nevertheless, this property can be regarded as favorable for HICs formation. HICs of halides of second group elements can serve as precursors for further producing of HICs with such important for applications insoluble compounds as ZnO, ZnS and CdS. Their comparative disadvantage is difficulty with anion remove in substitution reaction.

Computational modeling of HICs of PVA with halides (fluorides, chlorides, bromides and iodides) of second group elements from Be to Hg inclusive has been performed in Prosanov and Benassi [29]. We used nearly six molecular units for structure modeling and we believe that obtained local atomic conformation approximately represents repeating unit of whole structure. It was found that all these halides except mercury chloride and iodide are inclining to polymeric chain formation. It means that preliminary lined set of halide molecules reorganizes into polymeric chain after optimization. In such chain each metal ion is located in a center of distorted tetragon of halide ions as it is depicted in Figure 8.1.

For mercury chloride and iodide we failed to find initial arrangement leading to polymeric chain formation as result of optimization. Bulk structures were formed in these cases. Then, we calculated bonding energy per one molecular unit for polymeric form and compared it with earlier published enthalpy of crystal formation and calculated energy of dimerization of hypothetic halide molecules. As it can be expected, the first value was lower than the latter two (neglecting of sign of a value). It means that polymeric form is a less stable form than the crystalline one. It also means that dimerization which is a first step of crystallization is more preferable way than polymerization. These findings are in accordance with real situation and therefore justify our calculations. Exceptions are cadmium chloride, bromide and iodide which have weak deviations. For example, for cadmium chloride calculated bonding energy of polymeric form is −0.67 eV and calculated energy of dimerization is −0.64 eV. For cadmium bromide these values are −0.59 eV and −0.57 eV correspondently. For cadmium iodide they are equal. For zinc bromide bonding energy for polymerization

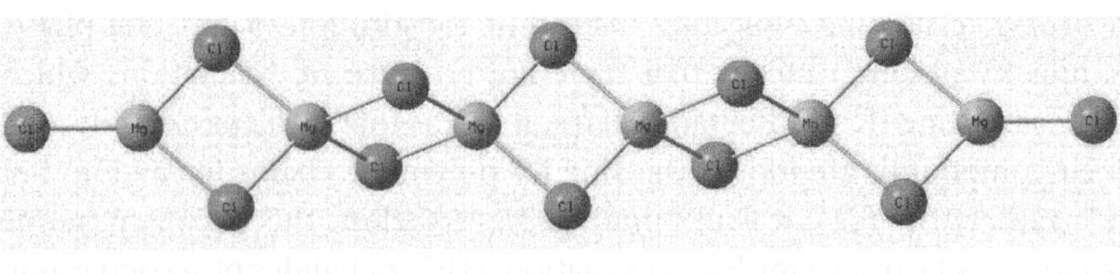

Figure 8.1 An example of optimized polymeric metastable structure of halides of second group elements.

also were found equal to energy of dimerization (−1.3 eV). It can be expected that metastable polymeric form of halides can be stabilized by complexing with PVA. Indeed, we found that some compounds ($BeCl_2$, $BeBr_2$, BeI_2, $MgCl_2$, MgI_2, $ZnCl_2$, and $ZnBr_2$) have bonding energy of single halide molecule with PVA higher than bonding energy for crystal structure. It testifies for preferable complex formation comparative to inorganic compound crystallization. And some compounds ($BeBr_2$, CaF_2, $CaBr_2$, ZnF_2, $ZnCl_2$, and $ZnBr_2$) have bonding energy of interpolymeric complex formation higher than enthalpy of crystalline state. Evidently, such compounds are especially suitable candidates for HICs production due to their thermodynamic stability.

Two distinct features of HICs were discovered from the modeling. 1) HICs have helix-like form evidently due to inconsistence of repeating units' lengths of PVA and inorganic components. For example for HICs of PVA with zinc fluoride, chloride, bromide and iodide computed periods of inorganic polymers are 2.9, 3.6, 3.6 and 4.2 angstrom correspondently. For HICs of PVA with cadmium halides these values are 3.4, 3.8, 3.9 and 4.3 angstrom. For PVA repeating period has been calculated 2.5 angstrom [30]. 2) There are two different forms of inorganic polymeric structures within HICs. One of them has two equal bridging halide ions. The second form has two different types of halide ions. One of them is bridging ion and the second is lateral ion. These two structures are represented in Figure 8.2.

They should have different shapes of Raman spectra. First structure should exhibit one band caused by "breathing" oscillation mode when

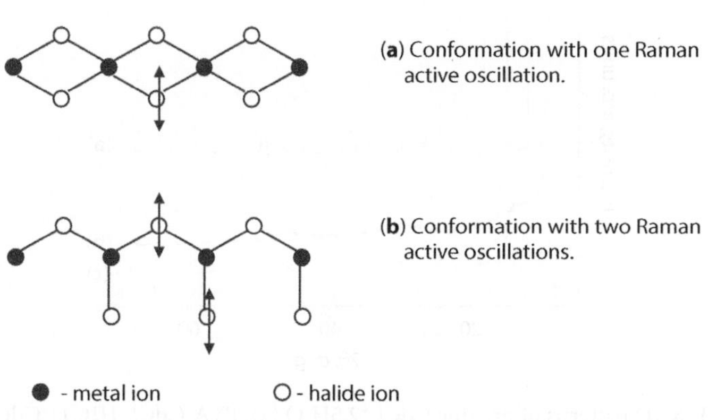

Figure 8.2 Types of inorganic polymeric structures of halides of second group elements and their Raman active oscillations.

two bridging halide ions of one structural unit moves toward and backward each other as it represented in Figure 8.2a. Second structure should exhibit two bands in Raman spectra (Figure 8.2b). One band is caused by "breathing" mode involving one bridging ion and another band is caused by valence oscillations of lateral halide ion. This prediction corresponds well with observations described below.

8.3.2.1 *Cadmium Halides Based HICs*

More thorough investigation of experimental and calculated properties of HICs of PVA with cadmium halides was undertaken in Prosanov *et al.* [30]. They represent difficult-to-dry plastic films. Typical XRD patterns of samples are represented in Figures 8.3 and 8.4.

According to XRD data, stoichiometric ratios of inorganic to organic components are 1 $CdCl_2$ molecule to 2 PVA repeating units and 1 $CdBr_2$ or CdI_2 molecule to 3 PVA repeating units for correspondent HICs. Computed structures exhibit nearly the same molar ratio of components as it is shown in Figure 8.5.

XRD patterns of PVA based HICs usually have a distinctive shape with two broad bands. In Prosanov *et al.* [30] this shape was attributed to incommensurable structures formation. In brief, in incommensurable structures two (or more) substructures with different periods can coexist.

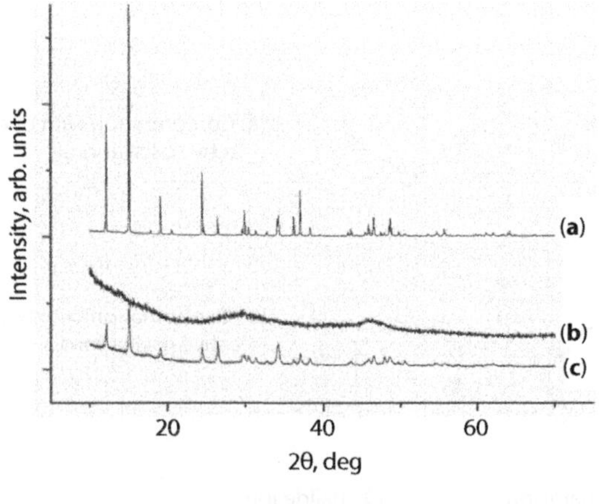

Figure 8.3 XRD patterns of pristine $CdCl_2$*$2.5H_2O$ (a), PVA-$CdCl_2$ HIC (1 $CdCl_2$ molecule to 2 PVA repeating units) (b) and PVA-$CdCl_2$ composition (1 $CdCl_2$ molecule to 1 PVA repeating units) (c).

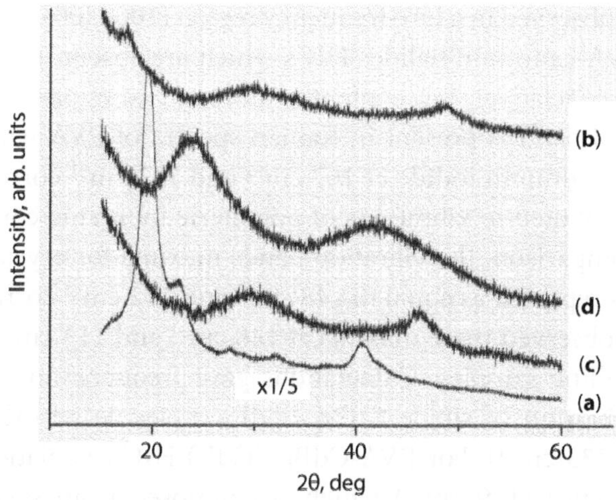

Figure 8.4 XRD patterns of PVA (a), PVA-$CdCl_2$ ($CdCl_2$/PVA units ratio = 1:2) (b), PVA-$CdBr_2$ ($CdBr_2$/PVA units ratio = 1:3) (c) and PVA-CdI_2 (CdI_2/PVA units ratio = 1:3) (d). Reproduced with permission from [30].

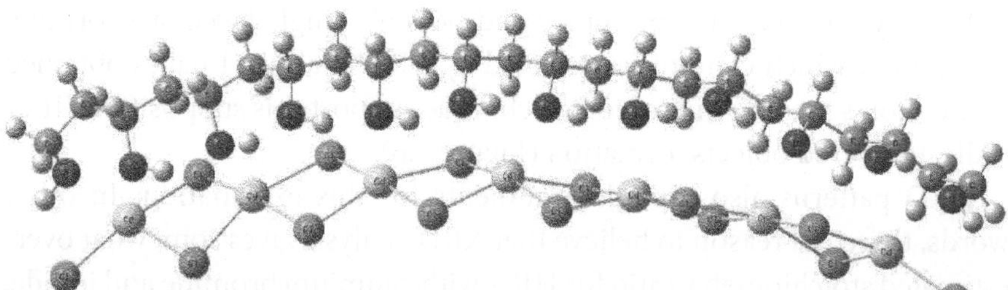

Figure 8.5 Computed structure of PVA-$CdCl_2$ HIC.

It is just a case of double helix of HICs where each of two spirals has its own period. A main diffraction peaks in XRD pattern of incommensurable structure have series of nearest weak satellites. One can believe that these series form unresolved broad bands in HICs XRD patterns. First band at lower diffraction angle corresponds to X-ray reflections over the planes parallel to the PVA chains. Another prominent XRD maximum at 40° for pristine PVA was attributed to diffraction on the planes perpendicular to PVA chains. This maximum approximately corresponds to an interplanar distance 2.2 angstrom which is close to the length of PVA repeating unit, *i.e.* 2.5 angstrom. At HICs XRD patterns both bands expectantly shift to lower diffraction angles with gradual increase of halide ion.

Bands are observed in a low-frequency region (100–300 cm^{-1}) of Raman spectra of PVA-cadmium halides HICs which are absent in the spectrum of pure PVA. At relatively low content of inorganic component in PVA only one additional band is present in Raman spectra of PVA with cadmium bromide and cadmium iodide at 167 cm^{-1} and 121 cm^{-1} correspondently. They were attributed to vibrations of polymeric inorganic components of HICs. For comparison, the vibration bands maxima for crystalline $CdBr_2$ and CdI_2 was reported as equal to 148 cm^{-1} and 112 cm^{-1} correspondently [31, 32]. We observed these maxima at 148 cm^{-1} and 113 cm^{-1} for the reference crystalline samples. Calculations gave frequency of symmetrical stretching vibration of single $CdBr_2$ (CdI_2) molecule adsorbed on PVA at 183 cm^{-1} (133 cm^{-1}). For PVA-$CdBr_2$ (CdI_2) HIC calculations gave the band at 163 cm^{-1} (119 cm^{-1}) which is composed from several normal modes of complex motions. With increase of halide concentration one more band appears in both cases at 144 cm^{-1} and at 113 cm^{-1}. And at the same time XRD patterns did not indicate presence of crystalline phase. It was explained by peculiarities of structure of cadmium bromide and iodide [30]. They are layered compounds and possibly single layers are forming in samples, which can not be detected by XRD analysis. Images obtained by electron scanning microscopy technique support this supposition. They indicate layered objects formation (Figure 8.6).

XRD patterns also provide arguments for this explanation. In other words, there is a reason to believe that XRD analysis gives somewhat overestimated stoichiometry ratio for HICs with cadmium bromide and iodide.

In the case of PVA-$CdCl_2$ HIC the situation is quite different. Only one band at 228 cm^{-1} was observed in the Raman spectra of the sample with stoichiometric components ratio determined by XRD analysis. With inorganic component content decrease this band slightly shifted to higher wavenumbers (to 231 cm^{-1} at ratio 1 $CdCl_2$ molecule to 20 PVA repeating units). Calculations gave the band at 279 cm^{-1} for single $CdCl_2$ molecule adsorbed on PVA and the band at 255 cm^{-1} compounded from several complex modes for PVA-$CdCl_2$ HIC. When content of chloride was higher than stoichiometric one several bands appeared in Raman spectrum of a sample. The same bands were observed in Raman spectrum of the reference crystalline sample of cadmium chloride hydrate $CdCl_2 \times 2.5\ H_2O$. Thus, in this case, we observe qualitative change in the Raman spectrum near stoichiometric components ratio, which supports our supposition concerning HIC formation.

Figure 8.6 Scanning electron images of PVA-CdI_2 HIC (1 CdI_2 molecule to 3 PVA repeating units) (a) and PVA-CdI_2 composition (1 CdI_2 molecule to 2 PVA repeating units) (b) at different resolution.

It was mentioned above that changes in HICs vibration spectra comparing with pristine PVA indicate interactions between HICs components. In IR spectra such changes usually occur in spectral range 2000 to 3600 cm^{-1} where O–H stretching modes are observed. Hydroxylic groups of PVA are interconnected via H bonding. This bonding shifts oscillation frequency. Due to PVA noncrystallinity quite wide range of such shifts results in high broadening of the absorption band. At HIC formation H bonding is substituted onto coordinative interaction of hydroxyls with metal ions. It leads to change of absorption band shape. Such change is clearly observed in spectra represented in Figure 8.7.

Changes of bands intensity and positions in the range 1000 to 1200 cm^{-1} in the HICs Raman spectra comparing to pristine PVA also indicate complexing. It is supposed that band of stretching oscillation of PVA's C–OH bond takes place there [33, 34]. Such changes are represented in Figure 8.8.

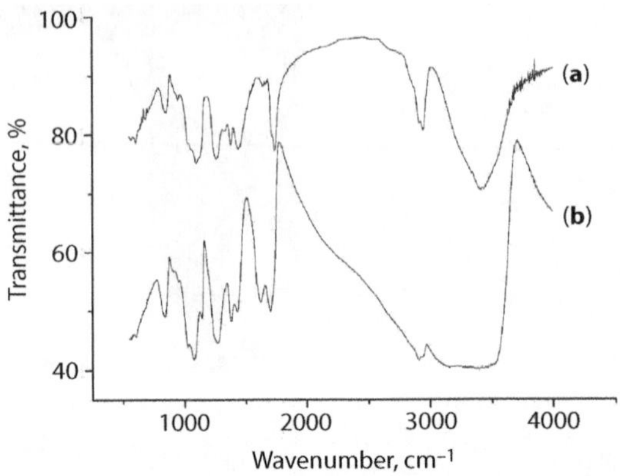

Figure 8.7 IR spectra of pristine PVA (a) and PVA–$CdCl_2$ complex (1 $CdCl_2$ molecule to 4 PVA repeating units) (b).

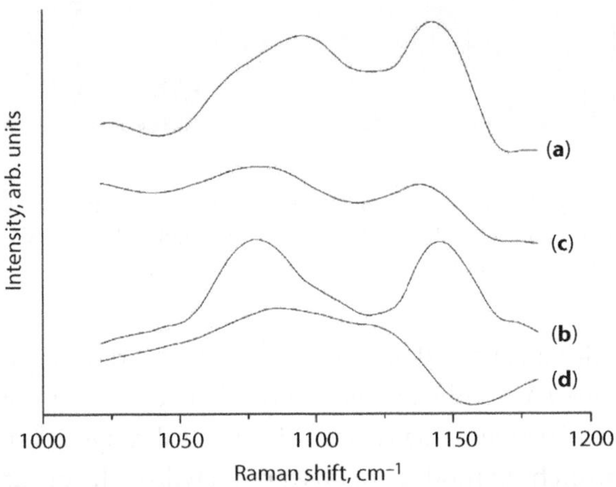

Figure 8.8 Raman spectra of pristine PVA (a), PVA-cadmium chloride, (b) PVA-cadmium bromide (c) and PVA-cadmium iodide (d) complexes in spectral range 1000 to 1200 cm^{-1}. Components ratio is 1 cadmium halide molecule to 4 PVA repeating units. Reproduced with permission from [30].

Calculated and observed HICs IR and Raman spectra in the range 2000 to 3600 cm^{-1} and 1000 to 1200 cm^{-1}, respectively ,satisfactory correlate each other.

Changes in UV spectra reflect modification of electronic structure at HICs formation. HICs with cadmium bromide and iodide turned out suitable materials for UV spectra investigations because they have absorption bands in region of wavelength between 200 and 300 nm. They can be considered as model systems which demonstrate general features for other HICs with adsorption bands lain beyond the bounds of commonly accessible spectral range. Two absorption bands present in the spectrum of HIC of PVA with cadmium iodide, whereas only one of them was found in the spectrum of cadmium iodide ethanolic solution (Figure 8.9). (Ethanol can be regarded as an analog of PVA with low molecular weight.)

Absorption band of PVA–CdI_2 composition at 220 nm is not sensitive to components ratio. This band also present in CdI_2 ethanolic solution. Another band is sensitive to concentration of cadmium iodide in PVA. It has maximum at 240 nm at components ratio 1 CdI_2 molecule to 1000 PVA units. At this ratio we believe that cadmium iodide forms simple complexes with PVA without HIC formation. At components ratio 1 CdI_2 molecule to 4 PVA units this band shifts to 265 nm. Evidently, HIC is formed in such mixture. The change in band position indicates interaction between neighbors cadmium iodide molecules adsorbed on PVA chain. In other words it indicates their polymerization. Calculations show that discussed optical transitions go between electronic states of inorganic component. PVA has no sufficient absorption in correspondent region.

It is known that analysis of edge of fundamental absorption band in solids gives information about their electronic structure [25]. In accordance with theoretical findings, the formula:

$$(E^*D)^r = f(E),$$

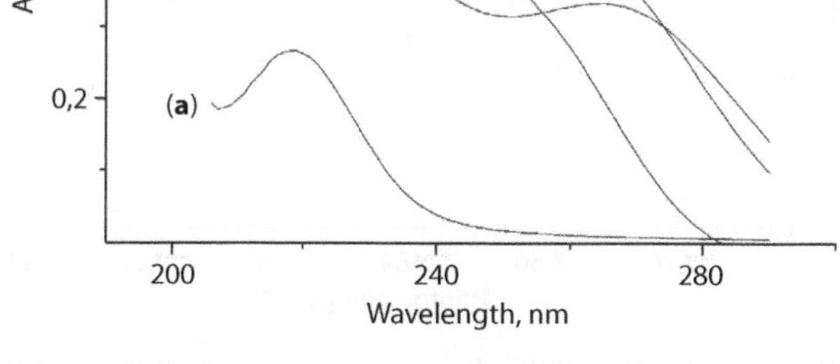

Figure 8.9 UV absorption spectra of CdI_2 in ethanol (a) and PVA-CdI_2 films. The $CdHal_2$/PVA units ratios are: 1:1000 (b), 8 (c), 4 (d). Reproduced with permission from [30].

where D is absorbance, should represent a linear dependence f(E) from photon energy – E at some value of exponent – r. This value indicates type of electronic transition. It can be one from numbers 1, 2/3, 1/2, 1/3 or 1/4 for crystalline structures. Noncrystalline solids often have another D(E) relationship (Urbach's low) [25]. Examples of such analysis concerning to PVA based systems are given in scientific publications. Direct allowed type of electronic transition was identified for to PVA–$TiCl_3$ complex in spite of its noncrystallinity [34]. A similar result has been reported for ZnO–CuO–PVA nanocomposition [35]. We performed approximations of the absorbance of PVA–cadmium iodide HIC in the ranges 3.76 to 3.94 eV (315–330 nm) and 3.50 to 3.60 eV (342–355 nm) and PVA-cadmium bromide HIC in the range 4.8 to 5.0 eV (250–260 nm). It was found out that in the former range it is better described by Gauss approximation, which is adopted for isolated optical centers [36] (Figure 8.10).

In the latter range Gauss's and Urbach's approximations were found approximately equally better than the expression for crystalline solids. The need in approximation in several spectral ranges arises from abrupt absorbance change. For PVA–cadmium bromide HIC UV absorbance approximations were carried out in the ranges 3.65 to 4.45 eV (280–340 nm) and 4.68 to 4.80 eV (260–265 nm). Gauss fit was preferable in both ranges. For PVA-cadmium chloride HIC UV absorbance approximations in the range

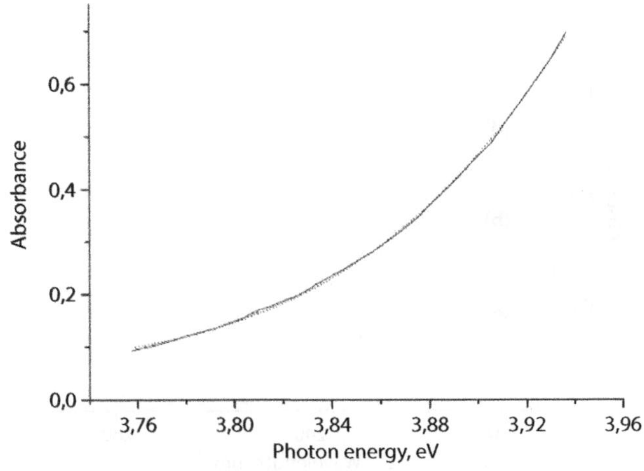

Figure 8.10 The approximation of the absorption spectrum (solid line) of PVA–cadmium iodide HIC with Gauss dependence (dot line).

3.65 to 5.20 eV (240–340 nm) is described approximately equally good with Gauss's and Urbach's fits. These results indicate structural difference of HICs from known nanocomposites.

8.3.2.2 *Zinc Halides Based HICs*

As it was mentioned in section 3.2, zinc chloride and bromide are regarded as highly expected candidates for HICs components due to thermodynamic stability of their complexes with PVA. Separate investigation of experimental and calculated properties of HICs of PVA with zinc halides is represented in Prosanov *et al.* [16]. Stoichiometric ratios were determined as 1 molecule of $ZnCl_2$ or $ZnBr_2$ to 1 repeating unit of PVA and 1 molecule of ZnI_2 to 4 repeating units of PVA. It corresponds to the calculated values. At the same time, zinc chloride and bromide are highly hydroscopic compounds and they also are known for glass forming ability [7, 37]. It means that overestimating of stoichiometric ratio based on XRD data can be admitted. Real samples of their HICs with PVA are quite plastic and difficult-to-dry.

XRD patterns of zinc halides and PVA HICs exhibit features similar to peculiarities of cadmium halides and PVA HICs. Namely, they contain two main wide bands which shift to low diffraction angle values at increasing of anion size. The same explanation of these patterns can be given as in the case of cadmium halides. Modeling predicts double-helix structure of PVA-zinc halides HICs.

Raman spectra of HICs of PVA with $ZnBr_2$ and ZnI_2 show two bands in a frequency range 100 to 400 cm^{-1} as it was predicted from calculations. In the case of PVA–$ZnCl_2$ HIC we observed one broad band which can be decomposed on two components. Calculated and observed positions of peaks are in satisfactory agreement. For example, for HICs of PVA with $ZnBr_2$ calculated values are 176 cm^{-1} and 199 cm^{-1} and observed ones are 182 cm^{-1} and 208 cm^{-1}.

8.3.3 Sulfides as HICs Components

As it was suggested in section 2, sulfides can be produced in PVA from appropriate HICs through solid state conversions. They tend to formation of usual compositions in PVA. Therefore, the existence of such

HICs is questionable in these systems for now. One more problem was revealed on this way. Halides based HICs are not suitable predecessors due to their high reaction stability. Sulfates were used in preliminary investigations of producing possibility of HICs with zinc, cadmium, nickel, manganese, cobalt and copper sulfides [19]. It is not known whether they form HICs or not. Stoichiometry of HICs was not determined in those experiments. Therefore, it would be better to denote used materials as "complexes" but not as HICs. Components ratio 1 sulfide molecule to 4 PVA residuals was used. Sulfate ion SO_4^{-2} has specific bands in IR and Raman spectra. The absence of these bands serves as indicator of complete sulfide formation. Saturated ethanolic solution of sodium sulfide was used as reagent. All initial PVA compositions with sulfates were amorphous according to XRD analysis. After treatment, crystalline phases of zinc, cadmium, manganese and cobalt sulfides were detected at XRD patterns of correspondent compositions. PVA with sulfides specimens were represented by brittle films.

"Dark" materials with cobalt, nickel and copper sulfides indicated featureless background in their Raman spectra. Therefore, this method cannot be used for indication of inorganic polymers formation in these cases. In Raman spectra of "clear" materials with zinc, manganese and cadmium sulfides bands of inorganic component were observed. They displayed only crystalline phases in two former compositions. In composition with cadmium sulfide two additional (to PVA) Raman bands at 300 cm^{-1} and 220 cm^{-1} were discovered. The first band was attributed to crystalline cadmium sulfide and the second one was attributed to polymeric cadmium sulfide. Calculations give a correspondent band at 260 cm^{-1}. The distortion was observed in 1000 to 1200 cm^{-1} spectral range of PVA–cadmium sulfide composition Raman spectrum similar to mentioned above changes in other HICs Raman spectra. Such distortion was also observed at Raman spectrum of PVA with cadmium sulfate, which was used as a precursor for PVA–cadmium sulfide synthesis.

Modelling of polymeric forms of sulfides of bivalence elements encounters problem with flat minimum of potential surface of molecular conformation. To overcome it we tried to find metastable cyclic conformations and succeeded in it for cycle of five zinc sulfide molecules. It was also found through calculations that sulfide of four-valence tin can form linear polymeric chain similar to that of second group elements halides depicted in Figure 8.1. Interaction with PVA stabilizes polymeric forms of sulfides.

As an example, optimized model of PVA–cadmium sulfide HIC is represented in Figure 8.11.

Terminal sulfur atoms in this model are substituted for univalent atoms (chlorine, for example) for the sake of system completeness. Optimized model of PVA–zink sulfide HIC has similar structure. Modelling indicated that PVA–SnS_2 HIC is unstable system, which decays on PVA and polymeric tin sulfide mixture. We failed to optimize PVA–NiS, PVA–CoS, and PVA–MnS HICs. We also failed to optimize PVA–CuS HIC but evidently, it was caused by mentioned above problem with potential surface flatness. Possible polymeric structure of copper oxide within HIC is quite different from that of cadmium and nickel sulfides. It is ribbon-like as it is represented in Figure 8.12.

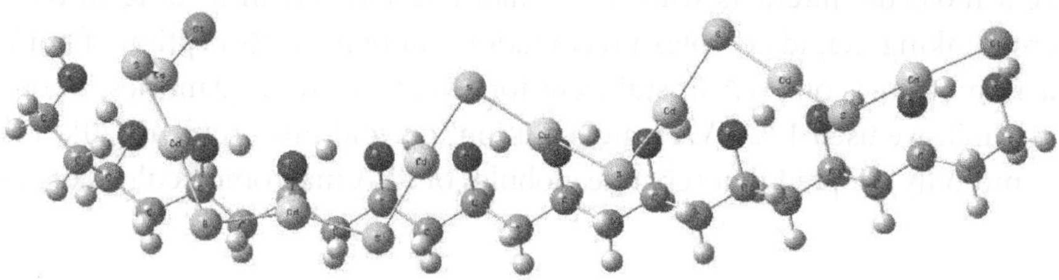

Figure 8.11 Optimized model of PVA–cadmium sulfide HIC.

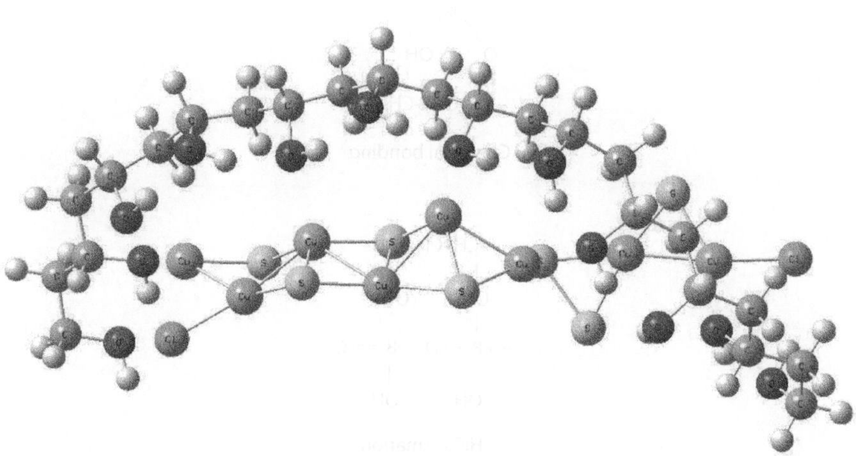

Figure 8.12 Possible structure of PVA-CuS HIC.

8.3.4 Boric Acid as HIC Component

Boric acid is quite attractive candidate as HICs component. A complex of PVA with (ortho-) boric acid $B(OH)_3$ have been investigated earlier [21, 38–40]. Boric acid belongs to group of compounds which interact with PVA quite strongly. Furthermore, it is one of few hydroxides which forms complex immediately. There is no need in soluble predecessor and intermediate reactions in this case. Therefore, HIC of PVA with boric acid can be regarded as model HIC system. However, such strong interaction has its own disadvantages. For instance, it leads to a rapid precipitation of complex from solution [21] with forming a substance hardening after drying. Two questions arise in this connection. 1) Whether boric acid and PVA form complex or chemically bonded compound? There are different opinions exist concerning this issue. Suggested structures of PVA complex with boric acid are representing at Figure 8.13. 2) It seems that each boric acid molecule interacts with more than one PVA chain leading to their crosslinking. Rapid complex precipitation can prevent absorption of boric acid molecules on PVA in sufficient for HIC formation quantities. In our research we used 1 % PVA water solution to avoid this problem [20]. It is commonly adopted that separate globules of PVA macromolecules exist in

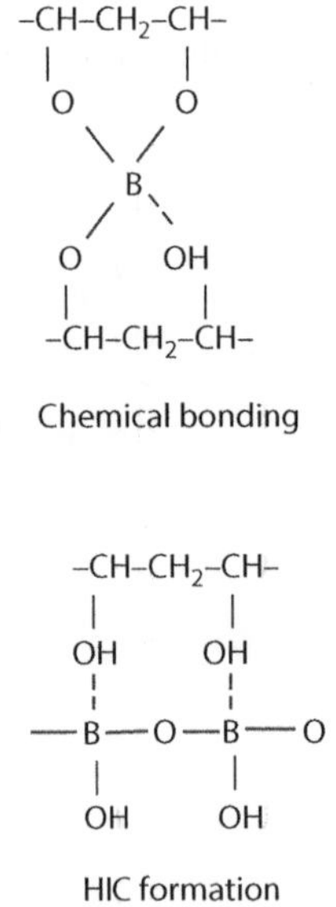

Figure 8.13 Suggested structures of PVA complex with boric acid.

solution with lower than 3% concentrations. Therefore, at lower concentrations interchains crosslinking can be excluded. But, intrachain crosslinking still remain. Such crosslinking complicates modeling sufficiently. In our research we neglected by this effect for the sake of computation model simplicity. HIC specimens look like brittle opaque films. At overstoichiometric content of boric acid its explicit precipitation becomes notable. It seems that oversaturated mixture tends to decay on HIC and boric acid instead of composition formation.

XRD patterns showed that stoichiometric ratio for HIC of PVA with boric acid is approximately 1:3. Calculations indicated that electronic bonding energy is positive for chemical bonding. Therefore, this model can be excluded as far as calculation reliability is adopted. The choice between complexing of boric acid in form of individual molecules and HIC formation remains. Calculations give nearly equal bonding energies per one boric acid unit (–0.37 and –0.36 eV). The difference between two structures consists in water content (in form of hydroxylic groups).

$$nB(OH)_3 \rightarrow (HBO_2)_n + nH_2O$$

Mass measurements rather give evidence for HIC formation. Calculated stoichiometry is quite different from one determined with XRD analysis. As it was mentioned above, this discrepancy can be explained by calculation inaccuracy or used model drawbacks. In contrast to cadmium and zinc halide based HICs modeling of boric acid based HIC does not exhibit distinct double helix formation.

Vibration spectra of PVA–boric acid complex, solid reagents and isolated boric acid molecules were also analyzed to substantiate HIC formation. The broad band of O–H bonds of H-bonded PVA hydroxylic groups at 3100 to 3600 cm^{-1} in IR absorbance spectrum is not changed significantly after boric acid addition into PVA in contrast with zinc and cadmium halide containing systems. The band at 2950 cm^{-1} is attributed to stretching C–H vibrations of PVA. The ratio of these bands' intensities remains approximately the same. It once again testifies for excluding of chemical interaction between PVA and boric acid. On the other hand, remarkable changes are observed in 1050 to 1150 cm^{-1} region of Raman spectrum where stretching vibrations of PVA C–O bonds and deformation vibrations of B–O–H bonds take place [33, 34]. It can be caused by complexation and H bonding of PVA and boric acid. In IR spectrum of supposed PVA–$B(OH)_3$ HIC two new bands appear at 770 cm^{-1} and 665 cm^{-1} as well. Raman spectrum of PVA–boric acid complex can approximately be represented as combination of bands of pristine PVA and two additional bands at 1250 cm^{-1} and 780 cm^{-1}. Calculated for HIC IR and Raman spectra were found in accordance with observations.

8.3.5 Copper Hydroxide/Oxide as HIC Component

PVA with copper hydroxide was the first substance which was suggested as genuine HIC forming [2] compound. PVA–copper hydroxide HIC can be obtained directly from ammonia solution, excluding predecessor conversion step. As it was mentioned in Introduction, PVA–copper oxide HIC formation is expected as a result of PVA–copper hydroxide HIC thermal decomposition. The latter is a subject of great interest due to diverse applications of copper oxide and it was a reason for investigating of these systems in Yu *et al.* [41]. Beneath we will resume CuO (not Cu_2O) as copper oxide.

Approximate stoichiometric ratio of $Cu(OH)_2$ to PVA in their HIC was determined as 1 $Cu(OH)_2$ molecular unit to 3 PVA repeating units according to XRD analysis. Calculated structures indicate stoichiometry close to 1:1. HIC specimens look like brittle green colored transparent films. At higher copper hydroxide content specimens become nonuniform. Opaque crumbly zone is forming in a center while border looks like stoichiometric HIC. Drying of a sample starts from a border and notable mass transfer to a center takes place. Similar to PVA–boric acid system, it seems that oversaturated mixture tends to decay on spatially separated areas of HIC and usual composition. HIC XRD pattern contains two broad bands inherent for other PVA based HICs. After vacuum heating during couple of hours at 420 K samples become black colored. Their XRD patterns indicate presence of unidentified reflection at $2\theta = 27°$ and absence of CuO peaks. Particles are observed on the specimen surface by means of electron scanning microscopy (Figure 8.14).

Such particles are absent at images of PVA–copper hydroxide HIC. There is no explanation of the nature of these particles for now. HIC weight decrease after heating approximately corresponds to expecting one.

Modeling revealed that copper hydroxide and oxide are able to the formation of metastable polymeric chains (Figure 8.15).

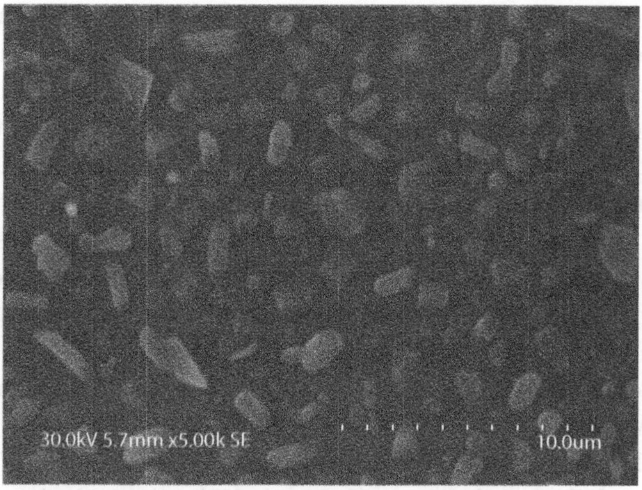

Figure 8.14 Scanning electron image of PVA–CuO HIC surface.

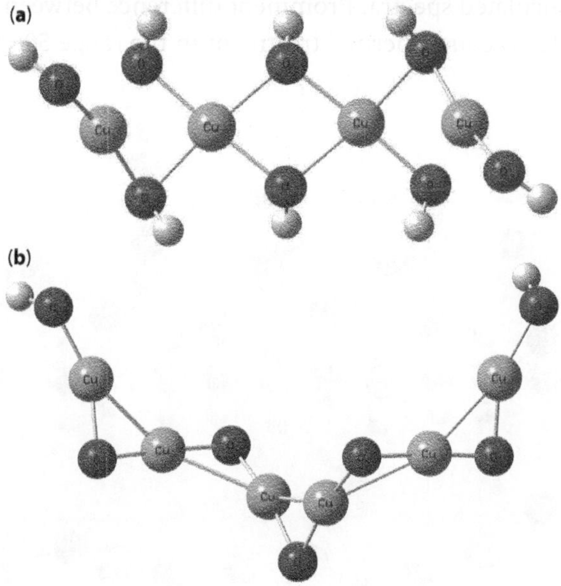

Figure 8.15 Examples of polymeric structures of copper hydroxide (a) and copper oxide (b).

Copper hydroxide can form several types of such chains. It is necessary to mention that chain structure can be observed in $Cu(OH)_2$ crystals as well [18, 42]. Calculations also indicate that copper oxide polymeric structure is variative. Two equilibrium Cu–Cu distances in the chain was discovered. This effect was not found in other investigated systems. Likely to boric acid–based HIC modeling of copper hydroxide and oxide-based HICs does not exhibit distinct double helix formation (Figure 8.16).

It also disaccords with existing suggestion [2]. The distinctive feature of calculated structures of PVA–copper oxide HIC is nonuniform Cu–Cu distance. Some Cu atoms tend to coupling with apparently randomly distribution within CuO polymeric chain. Calculated bonding energies of single copper hydroxide and oxide molecules absorbed on PVA (−2.0 and −4.8 eV correspondently) are sufficiently higher then bonding energies of HICs (−0.39 and −0.69 eV per one copper hydroxide/oxide unit).

Changes in IR spectra 3000 to 3700 cm^{-1} region are weak in comparison with PVA–cadmium halide HICs. Specific bands of crystalline $Cu(OH)_2$ are not observed in PVA–copper hydroxide HIC IR spectrum. The most distinctive difference between observed PVA–copper hydroxide/oxide HICs and pristine PVA spectra is the weak band at 600 cm^{-1}. It is not represented in calculated spectra. Prominent difference between HIC spectra before and after vacuum thermal treatment in the range 500 to 3000 cm^{-1}

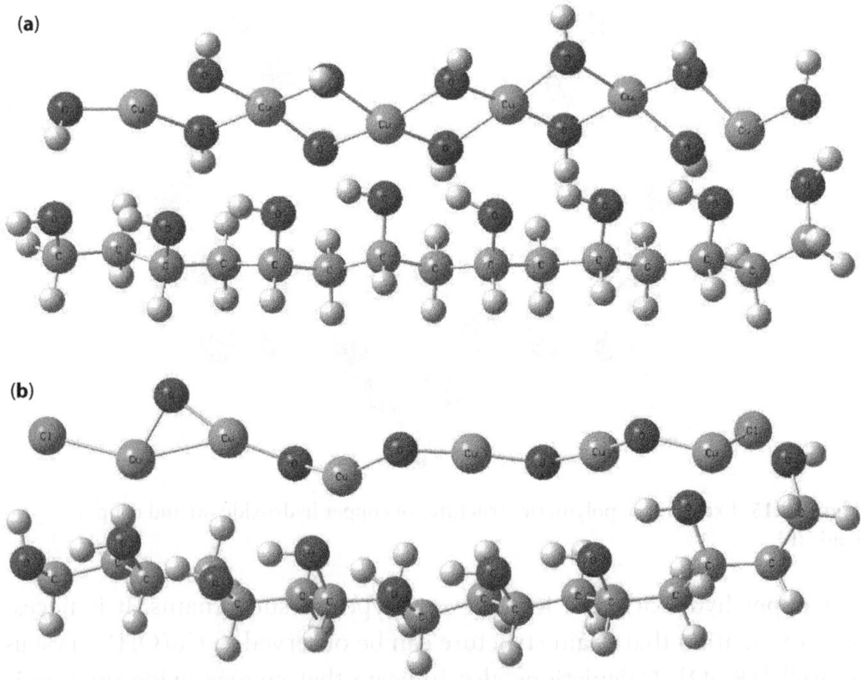

Figure 8.16 Models of PVA–copper hydroxide HIC (a) and PVA–copper oxide HIC (b).

was not found. Some changes of the broad band at 3000 to 3700 cm^{-1} were only observed. This effect was explained by supposition that copper hydroxide within HIC can exist in hydrated form. Our calculations confirm this supposition. Optimization of hydrated copper oxide structures results in hydrated mixed copper hydroxy-oxide species formation.

Raman spectroscopy was not informative method in PVA–copper hydroxide/oxide HICs investigation due to featureless background presence in spectra. Instead, UV spectroscopy gave valuable data. Two bands were found in UV-Vis region of absorption spectra of PVA–copper hydroxide films. The weaker band was observed at 650 nm. Nearly threefold variation of copper hydroxide content does not shift the position of this band. This band can be attributed to electronic transitions between copper d-orbitals split by electric field of the nearest ions. Evidently, insensibility of the position of this band to the copper hydroxide content is caused by insignificant participation of copper d-orbitals in chemical bonds formation. The position of the second band is affected noticeably by PVA to copper hydroxide ratio. A similar effect was described above in the section concerning to PVA–cadmium halides complexes. This band is observed at 242 nm in the spectrum of specimen with low copper hydroxide content (1 hydroxide molecule to 1000 PVA units) and at 261 nm in the spectrum of specimen with high copper hydroxide content (1 hydroxide molecule to 2 PVA units). After thermal treatment in vacuum, absorption in visible range increases gradually without developing any structure in a spectrum. In the UV region shift of absorption band to 252 nm in a spectrum of specimen with low copper hydroxide content and to 265 nm in a spectrum of specimen with high copper hydroxide content. It was supposed that isolated CuO molecules absorbed on PVA were formed in the first case and CuO polymeric chains were formed in the second case that results in different bands positions. Calculated spectra can be regarded in accordance with observations taking into account some suggestions. One of them is concerning to elemental bands broadening and overlapping. An important peculiarity of calculated PVA–CuO absorbance spectra is a presence of a strong band with maximum at 400 nm. This band is not present in observed spectrum. Instead of this band a continuous absorption is observed. Such behavior was attributed to a strong electron–phonon interaction, which results broadening of the band. Analysis of absorption spectral dependence exhibited that in range 3.6 to 4.8 eV (260–345 nm) it can be approximated best of all by Gauss law which corresponds to adsorption in isolated optical centers.

8.3.6 Hydroxides and Oxides Other then Copper Elements as HICs Components

Complexes of PVA with hydroxides and oxides of Be, Mg, Zn, Cd, Al, Cr(III) and Fe(III) were investigated in Prosanov *et al.* [1]. Complex of PVA with boric acid H_3BO_3 (which also can be regarded as hydroxide $B(OH)_3$) was investigated there as well. Producing of this complex excludes preliminary stage. It is the main reason why it was discussed separately. Data concerning complexes of PVA with titanic acid/titanium dioxide and with stannic acid/stannic dioxide were represented in Prosanov [43] and Prosanov *et al.* [44] correspondently. Stannic acid H_4SnO_4 and titanic acid H_4TiO_4 also can be regarded as hydroxides $Sn(OH)_4$ and $Ti(OH)_4$. Actually, there is a number of relative compounds like H_2SnO_3, $H_{10}Sn_5O_{15}$, H_3TiO_3, and so on. Their complexes have similarities with complexes of other hydroxides, which sometimes can be regarded as hydrated oxides. It is why they are discussed in this section. Only preliminary researches were carried out with materials mentioned above. Stoichiometry was not determined. Therefore, it is better to name the objects of investigations as "complexes" rather than HICs. Nevertheless, it seems that components ratio was not far from stoichiometric one, and obtained results can serve as a basis for more thorough researches. Precursors were produced from oligomeric beryllium hydroxide $BeCl_2$*$nBe(OH)_2$, magnesium sulfate $MgSO_4$*$7H_2O$, zinc bromide $ZnBr_2$, cadmium nitrate $Cd(NO_3)_2$*$4H_2O$, aluminium chloride $AlCl_3$, chromium sulfate $Cr_2(SO_4)_3$*$6H_2O$, iron chloride $FeCl_3$, titanium sulfate $Ti(SO_4)_2$, and tin chloride $SnCl_4$*$5H_2O$. Complexes with titanium compounds were prepared in components ratio 1 inorganic molecule to 3 PVA units. Other complexes were prepared in components ratio 1 inorganic molecule to 4 PVA units. Titanic, chromium and iron compounds precipitate PVA from solution due to quite strong interaction. Therefore, 1 % water PVA solution was used at producing of correspondent complexes according to the reason discussed in section 3.4. Water solutions of ammonium or potassium hydroxide were used for treatment of precursors. Hydroxides precipitated from water solutions of correspondent salts were used for comparison.

All specimens were produced in a form of brittle films. Specimens with hydroxides of beryllium, aluminum, iron and with stannic acid were quite transparent. Other specimens were rather opaque. XRD patterns of specimens with three- and four-valence metals hydroxides indicate absence of hydroxides peaks. Weak peaks of metal iron were observed at XRD pattern of specimen with iron hydroxide. In XRD patterns of specimens with all bivalence metals hydroxides weak peaks were present. XRD pattern of PVA–magnesium hydroxide composition exhibits peak of magnesium hydroxide. XRD pattern of PVA–zinc hydroxide composition exhibits peaks of zinc oxide. It is represented in Figure 8.17 as an example.

Unidentified peaks were present in XRD patterns of complexes with beryllium, cadmium hydroxides, and stannic acid. The presence of peaks on XRD patterns indicates that compositions are forming simultaneously with complexes. It is shortcoming of our method, which should be overcome.

Crystalline hydroxides of cadmium, magnesium and zinc display narrow absorption bands within 3000 to 3600 cm^{-1} IR region attributed to OH– groups valence vibrations. At their synthesis in PVA these bands disappear that indicates complexes formation. An example of such change is given in Figure 8.18.

Changes in 1000 to 1200 cm^{-1} region of Raman spectra where the band caused by PVA C–OH bond valence vibrations is placed indicates complexes formation. Such effect was observed for all considering systems. The absence of reference hydroxides Raman bands in spectra of their complexes with PVA reflects structural changes. This effect was observed for PVA with beryllium, cadmium, and zinc hydroxides. In Raman spectra of PVA complexes with chromium hydroxide, stannic, and titanium acids, the similar bands were found as for reference inorganic components. It means that structure of inorganic components

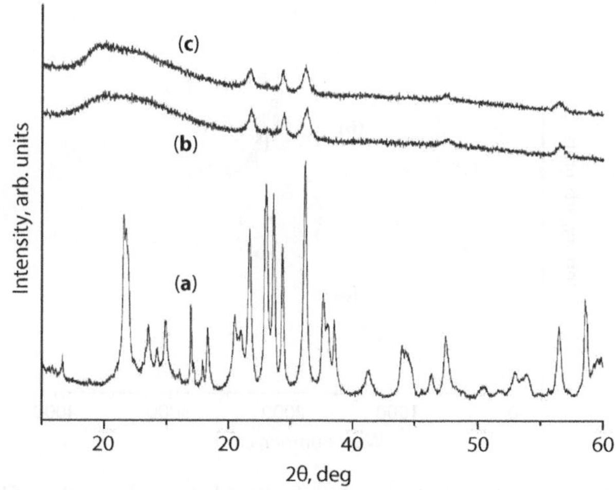

Figure 8.17 XRD patterns of reference zinc hydroxide (a) and complex of PVA–zinc hydroxide before (b) and after (c) vacuum heating.

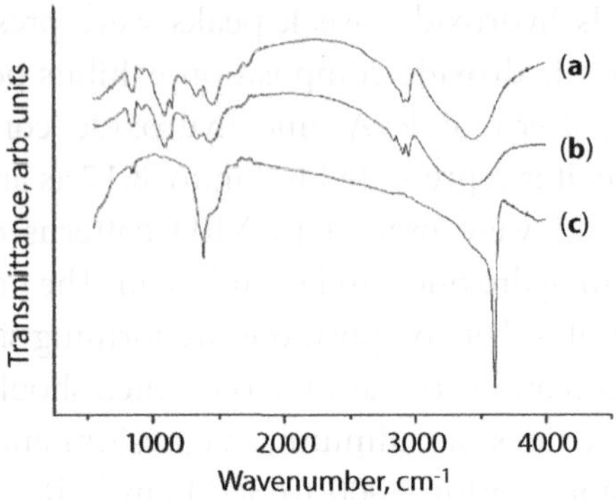

Figure 8.18 IR absorption spectra of pristine PVA (a), complex of PVA with cadmium hydroxide (b) and reference cadmium hydroxide (c).

is approximately the same in both cases. Appearance of specific bands in Raman spectra is regarded as indication of HIC formation. It was observed that such bands appear at compositions of PVA with aluminum hydroxide (after heating) (462 cm^{-1}), iron hydroxide (after aging) (330 cm^{-1}) and zinc hydroxide (178 cm^{-1}). An example of this effect is represented in Figure 8.19.

XRD and Raman scattering data indicate that PVA–magnesium hydroxide/oxide system has lowest intention to HIC formation among other systems discussed in this section.

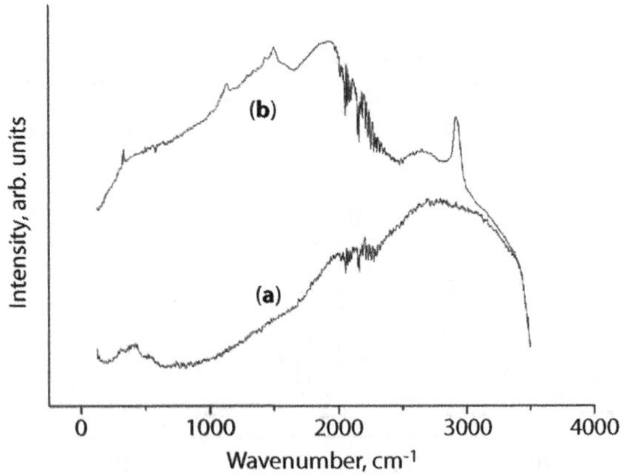

Figure 8.19 Raman spectra of reference iron hydroxide (a) and complex of PVA with iron hydroxide (b). Components ratio is 1 iron hydroxide molecule to 4 PVA repeating units.

It is supposed that complexes of PVA with oxides can be produced from complexes with hydroxides through thermal treatment. Decomposition of pristine PVA runs noticeably at temperatures higher than 200°C. Inorganic additives tend to decrease this limit. It restricts group of oxides which complexes with PVA can be produced by this way. In particularly, among considered compounds oxides of beryllium and magnesium are out of this threshold. Nevertheless, changes in IR spectrum can indicate polymeric beryllium oxide formation in dehydrated PVA which is a kind of polyacetylene [45]. Such system composed by two interpenetrated polymeric structures is an interesting material for optoelectronics. The formation of complexes with other oxides is also questionable. There is a good chance that it has zinc oxide due to low dehydration temperature of its hydroxide.

Modeling showed that some discussed oxides and hydroxides can exist in polymeric state and form HICs with PVA. Examples are represented in Figures 8.20 to 8.28. Evidently, they do not exhaust all structures. They rather illustrate conception.

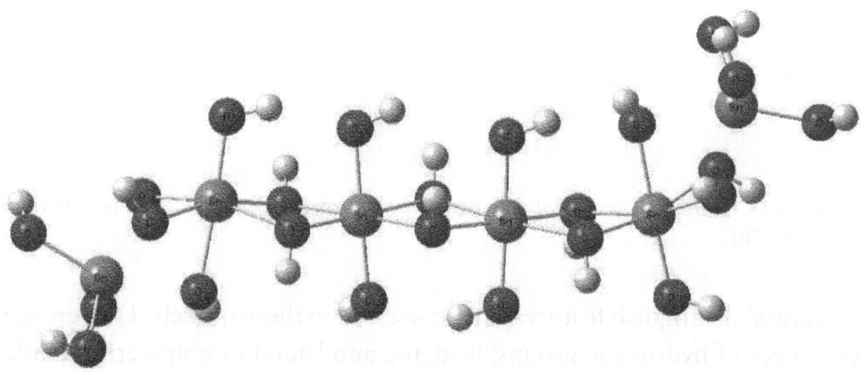

Figure 8.20 Optimized model of polymeric stannic acid.

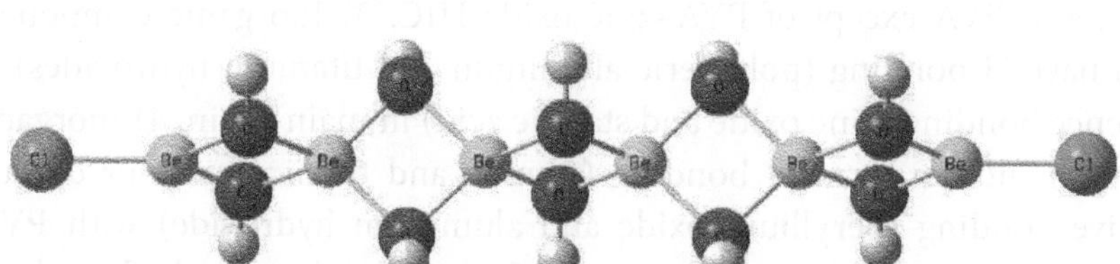

Figure 8.21 Optimized model of polymeric beryllium hydroxide.

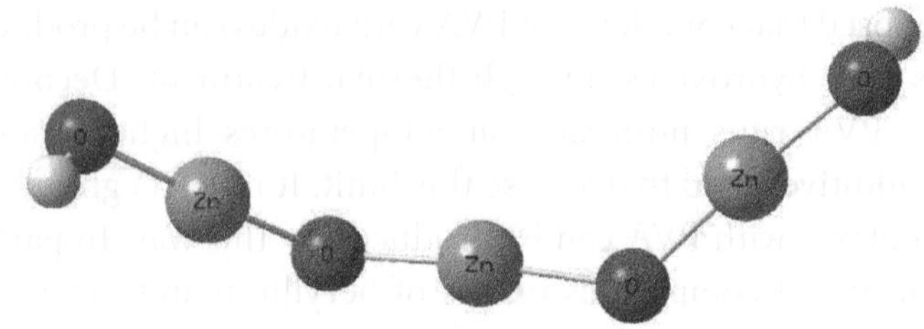

Figure 8.22 Optimized model of polymeric zinc oxide.

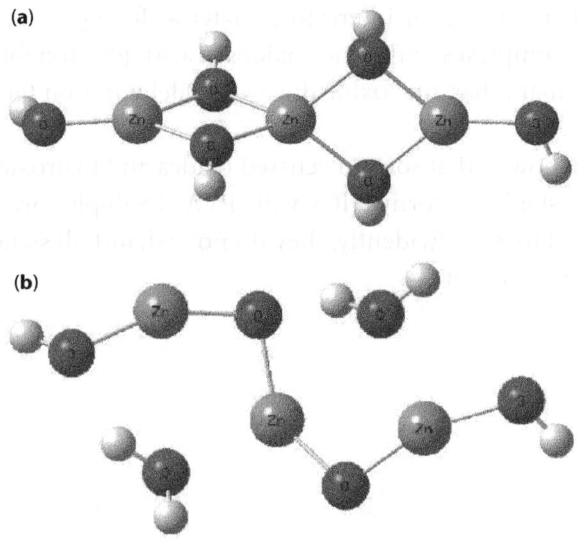

Figure 8.23 Optimized models of polymeric zinc hydroxide (a) and hydrated polymeric zinc oxide (b).

Several distinguish features can be stressed in these models. 1) There are two types of hydroxylic groups: bridging and lateral in polymeric stannic acid whereas only bridging hydroxyls exist in polymeric copper and zinc hydroxides. 2) There is no chemical bonding between inorganic component and PVA except of PVA–zinc oxide HIC. 3) Inorganic component can have H bonding (polymeric aluminum and titanium hydroxides) or valence bonding (zinc oxide and stannic acid) in main chain. 4) Inorganic component can form H bonding (stannic and titanic acids) or coordinative bonding (beryllium oxide and aluminum hydroxide) with PVA. 5) Inorganic component can exist in forms of polymeric hydroxide or polymeric hydrated oxide (zinc hydroxide).

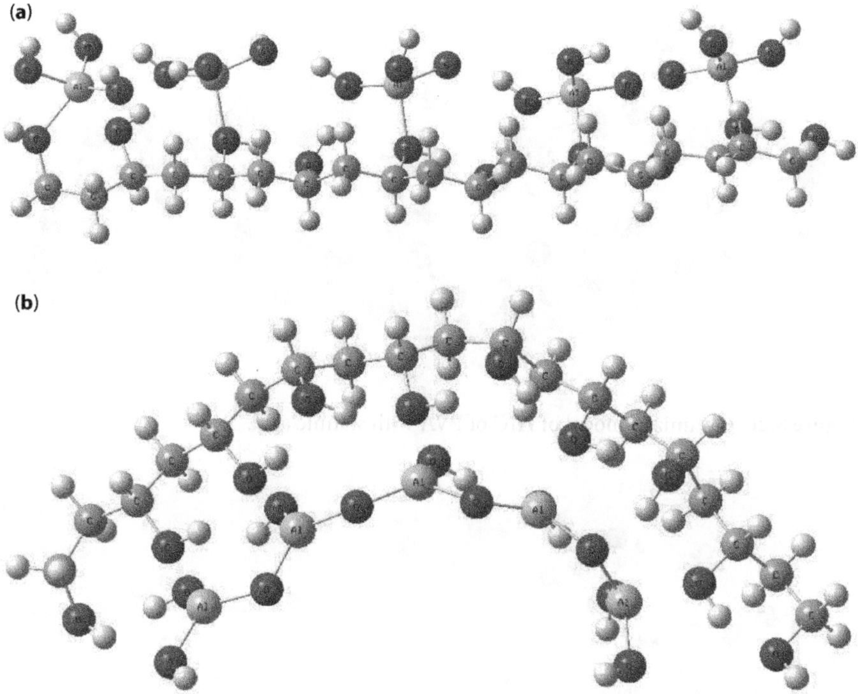

Figure 8.24 Optimized model of HIC of PVA with aluminum hydroxide (a) and partly dehydrated aluminum hydroxide (b).

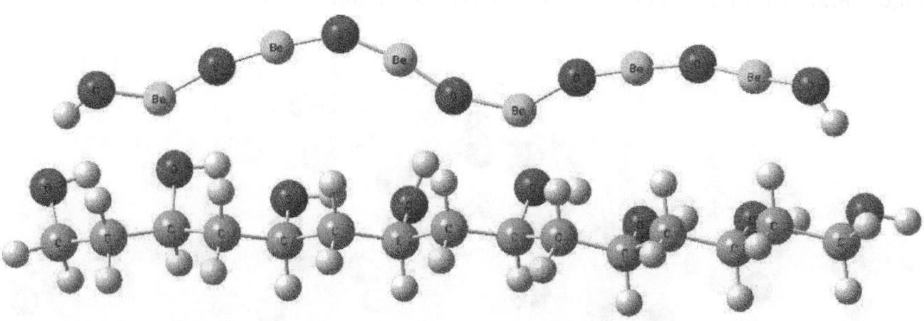

Figure 8.25 Optimized model of HIC of PVA with beryllium oxide.

Calculated Raman spectrum of PVA–aluminum hydroxide HIC exhibits distinctive band at 655 cm^{-1} attributed to symmetric vibrations of Al–OH bonds, which was not observed. It has been mentioned above that band at 462 cm^{-1} was observed at dehydrated PVA complex with aluminum hydroxide, but no prominent bands were found in its calculated spectrum. It seems, therefore, that calculations do not reflect investigated systems quite correctly.

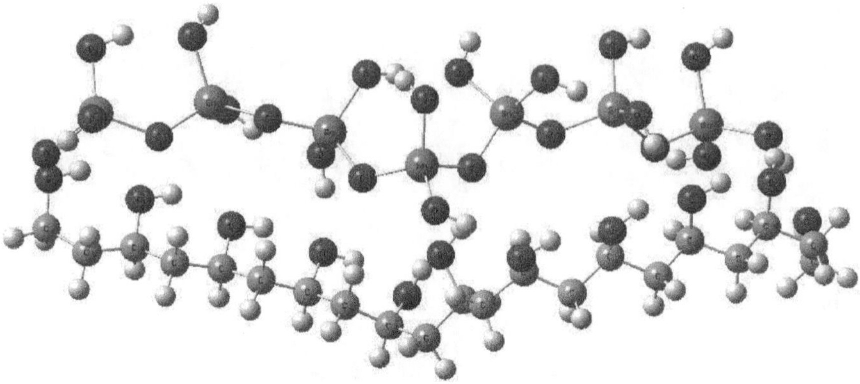

Figure 8.26 Optimized model of HIC of PVA with stannic acid.

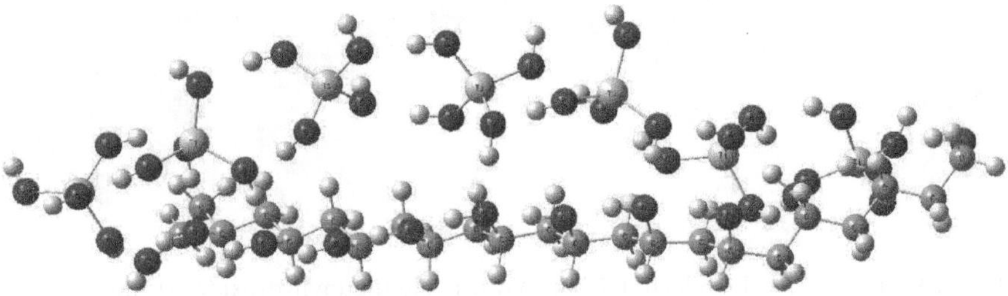

Figure 8.27 Optimized model of HIC of PVA with titanium acid.

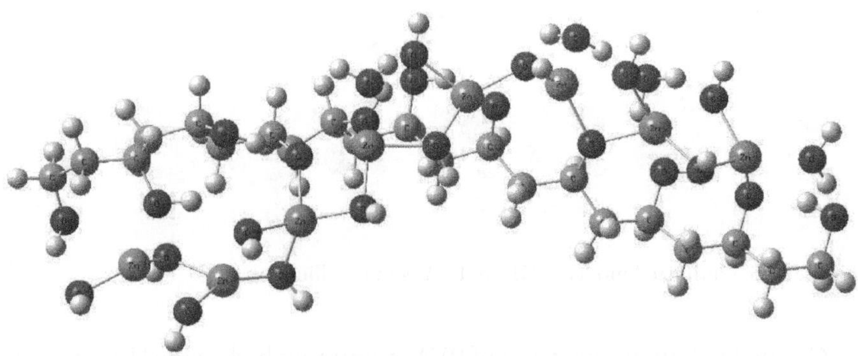

Figure 8.28 Optimized model of HIC of PVA with zinc hydroxide.

8.3.7 To Summarize

- Noncrystalline structure of HICs causes a need in complex of experimental methods and calculations for their investigation.
- Raman scattering is convenient technique for HICs structure determination.
- XRD analysis helps to find HICs stoichiometry.
- Zinc and cadmium halides are highly probable candidates as PVA-based HICs components.
- Producing of metastable HICs is not easily accomplishing task.
- Evidently, known complex of boric acid with PVA has a structure of HIC.
- HIC of PVA with copper oxide can be produced through thermal decomposition of HIC of PVA with copper hydroxide.
- There is a high probability of formation of PVA's HICs with hydroxides of tin, titanium, chromium, aluminum, beryllium, zinc and iron and their successful conversion into HICs with correspondent oxides or partly dehydrated hydroxides.

8.4 Possible Applications of HICs

Investigations of HICs are at the very beginning now. Therefore, there is not much data about their properties yet. However, it would be worth to estimate expecting benefits from their elaboration. For this purpose, it is reasonable to speculate how peculiarities of structure can influence properties of HICs as a kind of coordinative polymers. The most important applications of CPs discussed in R&D community are following:

1. photovoltaics,
2. optoelectronics,
3. batteries,
4. (gas) sensors,
5. membranes (gas separation),
6. gas storage,
7. drug delivery,

8. catalysis,
9. new functional materials.

Conducting behavior is crucial for the properties 1 to 4 and permittance for guest molecules is important for 3 to 8.

Review of CP's elaborations for photovoltaics application can be found in Xuanjun *et al.* and Rajnish *et al.* [46, 47]. Charge carriers are generated in photoelements under illumination. Then, they are being specially separated and collected with the aid of electrodes. Therefore, two processes determine an efficiency of photoelement. 1) Rate of intrinsic light conversion into electrical charge energy. 2) Efficiency of charge collection. For now, photoelements with p–n junction are dominating in photovoltaics. Structural heterogeneity is their distinctive feature. They provide quite high overall efficiency, but have some principle disadvantages. Among them are following: quite high production cost, narrow spectral range, and narrow active zone (restricted by p–n junction) where energy conversion takes place. The latter was suggested to overcome by the use of volume p–n junction. In such devices, p and n areas are dispersed and mixed in active volume at nanosize and even at molecular level. They are called as donor and acceptor components at the last case. This case is the most preferable for the process 1) accomplishment. MOFs are the very things where donors and acceptors (for electrons) are mixed at molecular level. They are mainly viewed as new materials for photovoltaics. However, they are not good candidates from viewpoint of process 2) because donors and acceptors do not constitute coherent structures in MOFs. In other words, donors (and acceptors) are separated from each other that hinder charge carriers collection in MOFs. This noncoherence is absent in nonporous CPs with 1D and 2D components in their structure, but suitable materials with quite high charge carriers mobility are unknown among them for now. It is probable that HICs became such compounds due to their higher potential for molecular design in comparison with common crystalline CPs [48]. This potential is stipulated by possibility of metastable structures formation. In optoelectronics the reverse aim according to photovoltaics is pursued. The efficient transformation of electrical energy into light is needed here. Once again, devices with p–n junction are dominating in this field. Their main disadvantages are other than that in photovoltaics. In particular, high transparency and flexibility of material are needed for some applications. These properties are inherent for polymers. Therefore, HICs are regarded as promising basic elements for optoelectronics.

Electronic conductivity is also important for battery electrodes and gas sensors design. The second crucial requirement here is possibility to guest ions or molecules hostage. In electrode and sensor materials outside ions or molecules correspondently should being reversibly incorporated into their structures. Therefore, these structures should not be very tight. MOFs and layered compounds together with nanocomposites are viewed as prospective in these fields. Electronic conductivity of electrode materials should be quite high, what is the problem for MOFs as it was mentioned above. In the case of sensors electronic conductivity can be low but it should depend sharply from the guest molecules incorporation. Pores in MOFs are rather large for outside ions used in charge/discharge processes and for some guest molecules in gas sensing applications. HICs have intermolecular space larger than that of crystalline CPs and smaller than MOFs have. Therefore, they can be regarded as candidates for electrode and sensing applications as well.

For membrane applications, selective absorption and permeability have key importance. Polymers are the most commonly used materials in this field. Their drawbacks are bad selectivity and permeability. They arise mainly from nonuniformity of intermolecular space and weak interaction of organic matrix with guest molecules. In contrast, MOFs and zeolites have uniform pores and inorganic substance, which interact rather strong with outside media components. Their main shortcoming is presence of intercrystallites boundaries. There is by-flow of separating mixture through these boundaries, which is difficult to prevent. HICs as a kind of polymers and CPs possess their properties favorable for membrane applications except pore uniformity. Instead of this, they can be subjected to different treatments influencing their porosity, which were elaborated in polymers technology.

Processes in gas storage materials are similar to those in gas sensors and membranes but desirable practical output is different. Here, reversible absorption of largest quantity of gases is a matter of interest. Correspondently, the ability of material to bind large amount of outside molecules becomes main issue. This bounding also should not be too strong. Storage of fuels like hydrogen and methane deserves greatest concern due to its practical benefits. Once again, MOFs are regarded as prospective materials in this field due to their high porosity and absorption ability. Zeolites should be mentioned among other effective gas storage materials. Clatrates are also under considerations. It is known that some metals can absorb hydrogen, but their capacitance is far away from acceptable condition. Hydrogen molecules are small and MOFs have too large pores for them. More dense structure of other CPs is more preferable for hydrogen uptake. Nonrigidity of HICs structure can give one more advantage for this purpose. We can appeal here to an example of polymeric hydrogels to illustrate a concept of gas storage by disordered systems. They are capable to high quantity of water absorption with increase of their volume. In our experiments, we observed reversible water absorption by PVA–zinc chloride HIC. Absorption was carried out at room temperature and 100% humidity, dehydration at 80°C. Reversible mass change was 20% in these conditions. Reference sample of silicagel did not exhibit noticeable mass change. It seems that HICs can serve as adjustable materials for water adsorption applications.

Drug delivery concept also has common features with gas storage. Here, drug molecules produce complexes with molecules of carrier. Then, these complexes are delivered to a target where drug molecules are released. This concept is relative to the concept of HICs formation, which was discussed in Part 2. Some drugs can exist in inactive poorly soluble forms. In opposite, their complexes can represent metastable active preparations. It is similar to the considered case of zinc (hydro)oxide and its complexes. On the other hand, polymers usually are not used as carriers in drug delivery and drug molecules do not interact each with other within a complex.

Catalytic materials work quite differently from ones discussed before. Here, active catalytic centers should interact with reagents. Therefore two following properties mainly determine overall process efficiency. 1) Existence of highly active catalytic centers. Usually, they are represented by chemical elements or compounds in highly dispersive form. Adsorption of catalytic centers on a carrier matrix prevents them from aggregation and deactivation. 2) High-contact surface should provide high rate of catalytic conversion. Therefore, carriers with high specific surfaces like zeolites or silica gel are usually used. MOFs have rather small pores to be a good carrier. On the other hand, the presence of moleculary dispersed inorganic components, which can serve as catalytic centers, makes them suitable for catalytic applications. HICs also have such highly dispersive active centers and as it has been mentioned above different methods of treatments are known influencing porosity of polymeric materials. It allows considering

HICs as potential catalysts. Antibacterial activity can be viewed as a kind of catalytic effect. Copper oxide is known for such activity, and it has been reported that PVA–CuO HIC exhibits good achievements as antibacterial preparation [49].

Peculiarities of HICs structure can condition their use as functional materials. One interesting field is use as binder in composites. It would be desirable, for example, to produce conducting compositions. Conductive materials like metals, graphite and (indium)-tin oxide (ITO) can have no properties like transparency or flexibility needed for special applications. Their compositions with HICs possibly can solve this problem. In particular, it would be worth to investigate conductivity of PVA–CuO HIC composition with CuO-based high-temperature superconductors because both components contain 1D CuO chains, which are regarded responsible for superconductivity. Other example of HICs use as functional materials has been suggested in Prosanov *et al.* [20]. Materials for neutron shielding usually contain atoms like boron with high cross-section of slow neutron capture. They also should include light atoms like hydrogen for neutrons deceleration. It should be optimal proportion between both atoms and they should be in atomic dispersive form in material efficient in neutron shielding. Composites and solutions of boron compounds usually are used in this field and increase their efficiency is not easy achieved goal. It has been predicted that neutron shielding efficiency of PVA-$B(OH)_3$ HIC is higher than that for known analogs. The similar material can be also used for neutron capture cancer therapy.

8.4.1 To Summarize

- HICs applications are not widely elaborated yet. They are rather hypothetical now.
- In general, they are similar to those for CPs
- The main factors determining HICs applications are as follows:
  - High dispersiveness of inorganic component
  - Its coherence (one dimensionality)
  - HICs' metastability
  - Intermolecular space/controlling porosity
  - Noncrystallinity

8.5 Conclusion

Concept of HICs is quite new and developing item in material science. It needs in recruiting of number of experimental methods and calculations for it substantiation. This conception was born from suggestions concerning structure of couple of known hybrid polymeric complexes and from ideas of growing crystals shape control through complexation with organic polymers. Its most attractive aspect is a way of production of metastable compounds with highly dispersed components that leads to valuable physical and chemical properties attainment. Applications of HICs are expected to be similar to CPs ones. Finally, the validation of HICs concept will be based on its fruitfulness for new materials elaboration.

References

1. Prosanov, I.Y., Bulina, N.V., Gerasimov, K.B., Complexes of polyvinyl alcohol with insoluble inorganic compounds. *Phys. Solid State*, 55, 2132–35, 2013.
2. Yokoi, H., Kawata, S., Iwaizumi, M., Interaction modes between heavy metal ions and water-soluble polymers. 1. Spectroscopic and magnetic reexamination of the aqueous solutions of cupric ions and poly (vinyl alcohol). *J. Am. Chem. Soc.*, 108, 3358–3361, 1986.
3. Takahama, T., Saharin, S.M., Tashiro, K., Details of the intermolecular interactions in poly(vinyl alcohol)-iodine complexes as studied by quantum chemical calculations. *Polymer*, 99, 566–579, 2016.
4. Sang, W., Fang, Y., Fan, J., He, Y., Min, J., Qian, Y., Novel synthesis method of ZnO nanorods by ion complex transformed PVA-assisted nucleation. *J. Cryst. Growth*, 299, 272–276, 2007.
5. Beilar Jr., J.C., Coordinative polymers, in: *Preparative Inorganic Reactions*, vol. 1, W.M. Jolly (Ed.), Interscience Publishers a division of John Wiley & Sons, New York, London, Sydney, 1964.
6. West, A.R., *Solid State Chemistry and Its Applications*, John Wiley and Sons, Chichester, 1984.
7. Greaves, G.N. and Sen, S., Inorganic glasses, glass-forming liquids and amorphizing solids. *Adv. Phys.*, 56, 1, 1–166, 2007.
8. Ginsberg, D.M. (Ed.), *Physical Properties of High Temperature Superconductors*, World Scientific, Singapore, New Jersey, London, Hong Kong, 1989.
9. Amo-Ochoa, P. and Zamora, F., Coordination polymers with nucleobases: From structural aspects to potential applications. *Coord. Chem. Rev.*, 276, 34–58, 2014.

10. Pettinari, C., Tabacaru, A., Galli, S., Coordination polymers and metal-organic frameworks based on poly(pyrazole)-containing ligands. *Coord. Chem. Rev.*, 307, 1–31, 2016.
11. Manners, I., Polymer science with transition metals and main group elements: Towards functional, supramolecular inorganic polymeric materials. *J. Polym. Sci. A Polym. Chem.*, 40, 179–191, 2002.
12. Mihailovic, D., Inorganic molecular wires: Physical and functional properties of transition metal chalco-halide polymers. *Prog. Mater. Sci.*, 54, 309–350, 2009.
13. Hassan, J.J., Mahdi, M.A., Chin, C.W., Hassan, Z., Abu-Hassan, H., Microwave assisted chemical bathdeposition of verticallyaligned ZnO nanorods on a variety of substrates seeded by PVA–$Zn(OH)_2$ nanocomposites. *Appl. Surf. Sci.*, 258, 4467–4472, 2012.
14. Wang, M., Lian, Y., Wang, X., PPV/PVA/ZnO nanocomposite prepared by complex precursor method and its photovoltaic application. *Curr. Appl. Phys.*, 9, 189–194, 2009.
15. Wang, H., Lu, X., Zhao, Y., Wang, C., Preparation and characterization of ZnS:Cu/PVA composite nanofibers via electrospinning. *Mater. Lett.*, 60, 2480–2484, 2006.
16. Prosanov, I.Y., Benassi, E., Bulina, N.V., Structure of hybrid interpolymeric complexes of zinc halides with polyvinyl alcohol. *Curr. Appl. Polym. Sci.*, 2, 44–48, 2018.
17. Haiyan, M., Yuguang, L., Hong, Y., Qiang, L., Minghao, R., Xubi, G., Ying, G., Yinfeng, L., Determination of coordination parameters of Cd^{2+} and Zn^{2+} with polyvinyl alcohol and chitosan. *J. Macromol. Sci. B Phys.*, 50, 1, 132–143, 2010.
18. Cudennec, Y. and Lecerf, A., The transformation of Cu(OH)2 into CuO, revisited. *Solid State Sci.*, 5, 1471–1474, 2003.
19. Prosanov, I.Y. and Bulina, N.V., Polymeric sulfides CdS, CuS, and NiS in polyvinyl alcohol matrix. *Phys. Solid State*, 56, 1270–1272, 2014.
20. Prosanov, I.Y., Abdulrahman, S.T., Thomas, S., Bulina, N.V., Gerasimov, K.B., Complex of polyvinyl alcohol with boric acid: Structure and use. *Mater. Today Commun.*, 14, 77–81, 2018.
21. Finch, C.A., Some properties of polyvinyl alcohol and their possible applications, in: *Chemistry and Technology of Water-Soluble Polymers*, C.A. Finch, (Ed.), pp. 287–306, Springer Science+Business Media, New York, 1983.
22. Shmidt, F., Vernetzungsreaktionen mit Estern der Orthotitansäure und anderen Metallalkoholaten. *Angew. Chem.*, 64, 19-20, 536–538, 1952.
23. Sierka, M., Synergy between theory and experiment in structure resolution of low-dimensional oxides. *Prog. Surf. Sci.*, 85, 398–434, 2010.

24. Brodsky, M.H., Raman scattering in amorphous semiconductors, in: *Light Scattering in Solids 1. Introductory Concepts*, Second Corrected and Updated Edition, M. Cardona (Ed.), pp. 205–252, Springer-Verlag, Berlin Heidelberg GmbH, 1983.
25. Mott, N. and Davis, E., *Electronic Processes in Non-Crystalline Materials*, Oxford University Press, Oxford, 1971.
26. Becke, A.D., Density-functional thermochemistry. III. The role of exact exchange. *J. Chem. Phys.*, 98, 5648–5652, 1993.
27. Foresman, J.B., Head-Gordon, M., Pople, J.A., Frisch, M.J., Toward a systematic molecular orbital theory for excited states. *J. Phys. Chem.*, 96, 135–149, 1992.
28. Frisch, M.J., Trucks, G.W., Schlegel, H.B., Scuseria, G.E., Robb, M.A., Cheeseman, J.R., Scalmani, G., Barone, V., Mennucci, B., Petersson, G.A., Nakatsuji, H., Caricato, M., Li, X., Hratchian, H.P., Izmaylov, A.F., Bloino, J., Zheng, G., Sonnenberg, J.L., Hada, M., Ehara, M., Toyota, K., Fukuda, R., Hasegawa, J., Ishida, M., Nakajima, T., Honda, Y., Kitao, O., Nakai, H., Vreven, T., Montgomery, Jr., J.A., Peralta, J.E., Ogliaro, F., Bearpark, M., Heyd, J.J., Brothers, E., Kudin, K.N., Staroverov, V.N., Keith, T., Kobayashi, R., Normand, J., Raghavachari, K., Rendell, A., Burant, J.C., Iyengar, S.S., Tomasi, J., Cossi, M., Rega, N., Millam, J.M., Klene, M., Knox, J.E., Cross, J.B., Bakken, V., Adamo, C., Jaramillo, J., Gomperts, R., Stratmann, R.E., Yazyev, O., Austin, A.J., Cammi, R., Pomelli, C., Ochterski, J.W., Martin, R.L., Morokuma, K., Zakrzewski, V.G., Voth, G.A., Salvador, P., Dannenberg, J.J., Dapprich, S., Daniels, A.D., Farkas, O., Foresman, J.B., Ortiz, J.V., Cioslowski, J., Fox, D.J., *Gaussian 09, Revision D.01*, Gaussian, Inc., Wallingford CT, 2013.
29. Prosanov, I.Y. and Benassi, E., Structure of hybrid interpolymeric complexes of polyvinyl alcohol and halides of second group elements. *Adv. Mater. Sci. Eng.*, Article ID 4931082, 2017.
30. Prosanov, I.Y., Benassi, E., Bulina, N.V., Matvienko, A.A., Structure and properties of self-assembling low-dimensional hybrid materials: The case of cadmium halides in polyvinyl alcohol. *Curr. Inorg. Chem.*, 7, 3, 155–161, 2017.
31. Anderson, A., Lo, Y.W., Todoeschuck, J.P., Raman and infrared spectra of crystals with the cadmium chloride structure. *Spectr. Lett.*, 14, 2, 105–116, 1981.
32. Anderson, A. and Lo, Y.W., Raman and infrared spectra of crystals with the cadmium iodide structure. *Spectr. Lett.*, 14, 8-9, 603–615, 1981.
33. Kuptsov, A.K. and Zhizhin, G.N., *Handbook of Fourier Transform Raman and Infrared Spectra of Polymers*, Elsevier, Amsterdam, Fizmatlit, Moscow, 19982001.

34. Abdelaziz, M. and Ghannam, M.M., Influence of titanium chloride addition on the optical and dielectric properties of PVA films. *Physica B*, 405, 958–964, 2010.
35. Abdullah, H., Azmy, N.A.N., Naim, N.M., Hamid, A.A., Idris, S., Synthesis and fabrication of ZnO–CuO doped PVA and ZnO–PbO doped PVA nanocomposite films by using γ-radiolysis and it's microbial sensor application. *J. Sol-Gel Sci. Technol.*, 74, 1, 15–23, 2015.
36. Curie, D., *Luminescence in Crystals*, Methuen & Co Ltd, London, John Wiley & Sons Inc., N-Y, 1963.
37. Yannopoulos, S.N., Kalampounias, A.G., Chrissanthopoulos, A., Papatheodorou, G.N., Temperature induced changes on the structure and the dynamics of the "tetrahedral" glasses and melts of $ZnCl_2$ and $ZnBr_2$. *J. Chem. Phys.*, 118, 3197–214, 2003.
38. Şenel, M., Bozkurt, A., Baykal, A., An investigation of the proton conductivities of hydrated poly(vinyl alcohol)/boric acid complex electrolytes. *Ionics*, 13, 263–266, 2007.
39. Sahin, A. and Ar, I., Synthesis, characterization and fuel cell performance tests of boric acid and boron phosphate doped, sulphonated and phosphonated poly(vinyl alcohol) based composite membranes. *J. Power Sources*, 288, 426–433, 2015.
40. Wang, N., Zhao, L., Zhang, C., Li, L., Water states and thermal processability of boric acid modified poly (vinyl alcohol). *J. Appl. Polym. Sci.*, 133, 13, 43246, 2016.
41. Prosanov, I.Y., Benassi, E., Bulina, N.V., Matvienko, A.A., Gerasimov, K.B., Sidelnikov, A.A., Polymeric copper oxide: Preparation and investigation of its structure and optical properties. *J. Inorg. Organomet. Polym. Mater.*, 28, 6, 2328–2335 , 2018.
42. Oswald, H.R., Reller, A., Schmalle, H.W., Dubler, E., Structure of copper(II) hydroxide, Cu(OH)2. *Acta Crystallogr. C Struct. Chem.*, 46, 2279–2284, 1990.
43. Prosanov, I.Y., Complex compound polyvinyl alcohol-titanic acid/titanium dioxide. *Phys. Solid State*, 55, 450–453, 2013.
44. Prosanov, I.Y., Chesalov, Y.A., Gerasimov, K.B., Bulina, N.V., Hybrid material polyvinyl alcohol-stannic acid/stannic oxide. *Phys. Solid State*, 55, 1519–1524, 2013.
45. Prosanov, I.Y. and Matvienko, A.A., Study of PVA thermal destruction by means of IR and Raman spectroscopy. *Phys. Solid State*, 52, 10, 2203–2206, 2010.
46. Zhang, X., Wang, W., Hu, Z., Wang, G., Uvdal, K., Coordination polymers for energy transfer: Preparations, properties, sensing applications, and perspectives. *Coord. Chem. Rev.*, 284, 206–235, 2015.

47. Kaur, R., Kim, K.-H., Paul, A.K., Deep, A., Recent advances in the photovoltaic applications of coordination polymers and metal organic frameworks. *J. Mater. Chem. A*, 4, 3991–4002, 2016.
48. Semenova, O.I., Yudanova, E.S., Yeryukov, N.A., Zhivodkov, Y.A., Shamirzaev, T.S., Maximovskiy, E.A., Gromilov, S.A., Troitskaia, I.B., Perovskite $CH_3NH_3PbI_3$ crystals and films. Synthesis and characterization. *J. Cryst. Growth*, 462, 45–49, 2017.
49. Hajipour, A.R., Mohammadsaleh, F., Sabzalian, M.R., Copper-containing polyvinyl alcohol composite systems: Preparation, characterization and biological activity. *J. Phys. Chem. Solids*, 83, 96–103, 2015.

Index

Also of Interest

Other Scrivener Publishing books by Visakh P.M.

Metal Oxide Nanocomposites
Synthesis and Applications
Edited by B. Raneesh and Visakh. P. M.
Published 2021. ISBN 978-1-119-36357-6

High Performance Polymers and Their Nanocomposites
Edited by Visakh P.M. and Semkin A. O.
Published 2019. ISBN 978-1-119-36365-1

Polypropylene-Based Biocomposites and Bionanocomposites
Edited by Visakh P.M and Matheus Poletto
Published 2018. ISBN 978-1-119-28356-0

Soy Protein-Based Blends, Composites and Nanocomposites
Edited by Visakh P.M. and Olga Nazarenko
Published 2017. ISBN 978-1-119-41830-6

Polyethylene-Based Biocomposites and Bionancomposites
Edited by Visakh P.M. and Sigrid Lüftl
Published 2016. ISBN 978-11903845-0

Nanostructured Polymer Membranes
Volume 1: Processing and Characterization
Edited by Visakh P. M. and Olga Nazarenko
Published 2016. ISBN: 978-1-118-83173-1

Nanostructured Polymer Membranes
Volume 2: Applications
Edited by Visakh P. M. and Olga Nazarenko
Published 2016. ISBN 978-1-118-83178-6

Polyethylene-Based Blends, Composites and Nanocomposites
Edited by Visakh P. M. and María José Martínez Morlanes
Published 2015. ISBN 978-1-118-83128-1

Polyoxymethylene Handbook
Structure, Properties, Applications and Their Nanocomposites
Edited by Sigrid Lüftl, Visakh P.M., and Sarath Chandran
Published 2014. ISBN: 978-1-118-38511-1

Advances in Food Science and Technology
In two volumes
Edited by Visakh P.M., Sabu Thomas, Laura B. Iturriaga, and Pablo Daniel Ribotta
Published 2013. 978-1-118-12102-3

Handbook of Engineering and Specialty Thermoplastics 4
Volume 4: Nylons
Edited by Sabu Thomas and Visakh P.M.
Published 2012. ISBN 978-0-470-63925-2

Handbook of Engineering and Specialty Thermoplastics 3
Volume 3: Polyethers and Polyesters
Edited by Sabu Thomas and Visakh P.M.
Published 2012. ISBN 978-0-470-63926-9

Printed and bound by CPI Group (UK) Ltd, Croydon, CR0 4YY
17/09/2026
14964245-0001